THE
INTELLIGIBLE
UNIVERSE

An Overview of
the Last Thirteen
Billion Years

2ND EDITION

THE
INTELLIGIBLE
UNIVERSE

An Overview of the Last Thirteen Billion Years

2ND EDITION

JULIO A. GONZALO

Universidad Autónoma de Madrid
Spain

 World Scientific

NEW JERSEY · LONDON · SINGAPORE · BEIJING · SHANGHAI · HONG KONG · TAIPEI · CHENNAI

Published by

World Scientific Publishing Co. Pte. Ltd.

5 Toh Tuck Link, Singapore 596224

USA office: 27 Warren Street, Suite 401-402, Hackensack, NJ 07601

UK office: 57 Shelton Street, Covent Garden, London WC2H 9HE

British Library Cataloguing-in-Publication Data
A catalogue record for this book is available from the British Library.

THE INTELLIGIBLE UNIVERSE (2nd Edition)
An Overview of the Last Thirteen Billion Years

Copyright © 2008 by World Scientific Publishing Co. Pte. Ltd.

ISBN-13 978-981-279-410-9
ISBN-10 981-279-410-7
ISBN-13 978-981-279-411-6 (pbk)
ISBN-10 981-279-411-5 (pbk)

Printed in Singapore by World Scientific Printers

To my brothers Bernardo, José, Manuel and Leopoldo, their wives and children.

To our parents, Leónides and Concepción, who are already in the presence of their Creator.

Contents

"The world must be some shape, and it must be that shape and no other; and it is not self-evident that nobody can possibly hit on the right one."

Gilbert Keith Chesterton,
"All is grist", pp 7–8
(New York: Dodd, Mead, 1932)

Prologue to 2nd Enlarged Edition

Except for the Second Section of this book (Chapters 7 to 10) in which the "New Inflationary Cosmology" was considered in some detail by distinguished invited speakers (John C. Matter and George F. Smoot) at the Summer Course at El Escorial August 16 to 19, 1993, the rest of it uses basically the standard <u>Friedmann–Lemaitre</u> Big Bang Cosmology (with k < 0, Λ = 0) to describe most of the available observational evidence. The question of the reported cosmic acceleration (S. Perlmutter, 1998) which, according to Martin Rees (Spring, 2000), makes the case for a non-zero cosmological constant strong but not overwhelming, is left aside for the time being.

In the First Section (Chapters 1 to 6) the masses, distances and times in the universe are discussed in the perspective of Relativistic Cosmology, among other things. The material is the same as the one included in the first edition of "The Intelligible Universe" (Shanghai Educational Publishing House, 1993).

In the Second Section the English version of two invited lectures given at El Escorial are reproduced. They were presented by John C. Mather (COBE's Project Scientist and FIRAS Principal Investigator) and George F. Smoot (COBE's DMR Principal investigator). The Spanish version of these two lectures was publised in book form in "Cosmología Astrofisíca" (Alianza Universidad: Madrid, 1994) eds. J. A. Gonzalo, J. L. Sánchez Gómez, M. A. Alario. Other distinguished speakers at El Escorial are given bellow:

Professors Ralph A. Alpher (New York), Antonio Fernandez Rañada (Madrid), Jose L. Sánchez Gómez (Madrid), Manuel Catalan Perez-Urquiola (Cadiz), Hans Elsässer (Heidelberg), Julio A. Gonzalo

(Madrid), Stanley L. Jaki (Princeton) author of "God and the Cosmologists" and J. Lejeune (Paris).

As it is well known Jonh C. Mather and Goerge F. Smoot received the 2006 Nobel Prizes in Physics for "their discovery of the blackbody form and anisotropy of the cosmic microwave background radiation". To say the least, these Nobel Prizes were not unexpected.

The Third Section ("The last thirteen billion years") summarizes the contents of a book published in Spanish by the UAM (Universidad Autónoma de Madrid) one year after the results given in a paper by N. Cereceda, G. Lifante and Julio A. Gonzalo in "Acta Cosmologica" (Krakow, 1998) anticipating accurately the time elapsed since the Big Bang as (13.7 ± 0.2) billion years.

The Fourth (very short) Section (from Chapter 6 of "Inflationary Cosmology Revisited" World Scientific, 2005) comments on NASA's WMAP result reporting the first full year accurate observations resulting in $t_0 = (13.7 \pm 0.2) \times 10^9$ years and $H_0 = R_0/R_0 = 67$ km/Mpcsec.

The Fifth Section presets a historical discusion of the medieval roots of contemporary science (physics and astronomy in particular) following the steps of P. Duhem and S. L. Jaki. Not so long ago, for many physicist chemists, mathematicians "the Wisdom of God is manifested in the works of his Creation."

And the (very short) last Section summarizes some important cosmic numbers before making some relevant concluding remarks.

In summary, the universe is well made by a Wise, and Allpowerful Creator, and man's intellect, limited as it is, also is well made by Him in His image and likeness. That is the reason why man can find the intelligible Universe after all.

Julio A. Gonzalo
Madrid, 19 March 2007
Festivity of Saint Joseph

Acknowledgements and Credits

I wish to thank my good friends and colleagues Fr. Manuel M. Carreira, SJ, Gines Lifante (UAM) and Manuel I. Marques (UAM) for reading the manuscript and helpful comments; Mª Felisa Martínez Ruiz for revising and typesetting the manuscript and helping to choose the illustrations; Prof. Ralp A. Alpher for helpful correspondence during 1994–1996 and Prof. Stanley L. Jaki for many instructive conversations during his a-periodic visits to Madrid since 1990 to present.

Credits for the illustrations and photos appearring with minor or major modifications in the various chapters of this book have been given whenever possible and useful. In particular, the web pages of NASA (www.nasa.gov) and ESA (www.esa.int/); Shanghai Educational Publishing House (Shanghai, 1993); Harper and Row (New York, 1969); Warner Books (New York, 1980); W. W. Norton & Co. (New York-London, 1978); Alianza Universidad (Madrid, 1995); Ediciones UAM (Madrid, 2002) and Mister Kerry Magruder for the pictures of Copernicus Galileo, Kepler and Tycho.

The Intelligible Universe
(Shanghai/Madrid 1993)

Chapter 1

Man and His Universe

1.1. Einstein's Eternal Mystery

Albert Einstein (1879–1955) made once a famous remark to the effect that the most surprising thing about the physical world is that it is intelligible. Writing to his friend Maurice Solovine[1] on March 30, 1952, i.e. few years before his death, Einstein said "You find it surprising that I think of the comprehensibility of the world (insofar as we are entitled to speak of such world) as a miracle or an eternal mystery. But, surely, *a priori*, one should expect the world to be chaotic, not to be grasped by thought in any way. One might (indeed should) expect that the world evidenced itself as lawful only so far as we grasp it in an orderly fashion. This would be a sort of order like the alphabetical order of words. On the other hand, the kind of order created, for example, by Newton's gravitational theory is of a very different character. Even if the axioms of the theory are posited by man, the success of such a procedure supposes in the objective world a high degree of order, which we are in no way entitled to expect *a priori*."

Man's intelligence has been confronted through all the millennia of recorded history with his personal reality as well as with the reality of the physical universe. Sometimes, especially in antiquity but also in modern times, man has viewed his personal reality as dissolved, so to say, in the hugh reality of the physical universe. Sometimes, on the contrary, he has seen the universe as a byproduct of his own mind. None of these two viewpoints is entirely satisfactory. The physical universe can be rationally defined as the totality of matter and energy (including of

3

course, electromagnetic radiation) in coherent interaction with one another. Intelligibility, in its deepest sense, involves a higher quality in man, the observer, a quality which is spiritual in nature. All cultures, all developed languages, seem to contain, however veiled, this distinction between the physical realm, dominated by brute inexorable forces, and the spiritual realm. Man's intelligence can discern -"read-into"- the physical world, and in the spiritual world of his fellowmen as well, truth beauty, unity, etc. A distinctive characteristic of man's intellectual operations is that they can go wrong, they can make mistakes, in clear contrast with the generally deterministic behavior of the material world.

1.2. From Antiquity to the XVI Century

A cursory overview of the historical record brings forth the well known fact of man's curiosity and interest on cosmic regularities from the earliest antiquity. Astronomy is said to be the oldest of pure sciences. Soon afterwards men's interests in the heavens were connected with practical matters, the first of which was to provide a basis for the calendar. The Chinese, Egyptians and Mayas had all very old working calendars, based on astronomic observations,[2] which helped them to carry out in due time their agricultural labors, to record past events and to calculate in advance dates for future plans.

The time needed for Earth to complete an orbit around the sun is 365 days, 5 hr, 48 min and 46 sec. On the other hand the moon passes through its phases in 29 days and 12 hr, approximately, and twelve lunar months amount to slightly more than 354 days, 8 hr and 48 min. Therefore defining an exactly overlapping solar and lunar year meets with inescapable difficulties. On one hand, months and years cannot be divided exactly into days. On the other, years cannot be easily divided into months. To reconcile and harmonize solar and lunar records has been a major problem for men of ancient civilizations, and different peoples have used different intercalation procedures.

In the year 45 B.C., Julius Caesar, who as "pontifex maximus" had the power of regulating the calendar, added 90 days to the year 46 B.C.

(making up for accumulated delays) and changed the years' length to 365 and 1/4 days, introducing therefore the bissextile year, on the advice of the astronomer Sosigenes. This was pretty close, but sixteen centuries latter the accumulated surplus time had displaced the vernal equinox from March 21 (the date set in the 4th century) to March 11. In 1582 Pope Gregory XIII rectified this error excepting from the condition of bissextile those years ending in hundreds, unless they are divisible by 400. Then the actual solar year equal to 365.2421 days, was approximated by

$$365 + 1/4 - 1/100 - 1/400 = 365.2425 \text{ days} \qquad (1.1)$$

which is close enough, so that correction of one day is required only after approximately 2500 years, a conveniently long time span.

The early astronomers of Babylonia, Assyria and Egypt were priests and a strong motivation for detailed study of the apparently irregular planetary motions (planet = "wanderer") was the preparation of horoscopes. An all pervasive trend[3] in those ancient cultures, as well as in the most brilliant of them all, the Greek culture, was that going down from the higher realm of the fixed stars, incorruptible and moving in perfectly orderly fashion, to the realm of the planets, less orderly and incorruptible, and finally to the sublunary world, the degree of disorder increased steadily, being the lower governed to a greater or lesser extent by the higher realms. This explains the need for horoscopes to predict the future and the mixed character (astronomical - astrological) of their investigations of the heavens.

Table 1.1 gives succinctly some of the major achievements concerning astronomy up to the times just before Galileo and Newton.

The pattern of the constellations or star groupings was inherited by Greeks and Egyptians from the oldest Mesopotamian culture, which, more than 5000 years ago, divided the band of stellar background near the plain defined by the apparent diurnal rotation of the sun around the Earth in twelve equally spaced arc segments.

Fig. 1.1. Apparent trajectory of the sun through the sky.

While some archeologists believe the original number[4] of zodiacal constellations was six, this may have been changed at later times, resulting in the twelve familiar constellations depicted in Fig. 1.1. In the resulting division of the zodiac by Mesopotamian astronomers-astrologers there is a definite tendency to duplicate things: two serpents, two slayers of serpents, two fishes, two streams and the twins making up Gemini. By the year 3000 B.C. the easily recognizable constellations Taurus and Scorpio were at the two opposite equinoctial points (day and night of equal duration) corresponding to the days marking the beginning of spring (vernal equinox) and the beginning of fall (autumnal equinox).

An observer watching at sunrise in the vernal equinox could see the position of the sun in the zodiac at Taurus by the year 3000 B.C. Two thousand years later, however, Taurus had moved one step and what the observer could see at sunrise in the vernal equinox was the constellation of Aries. And so forth. Towards the end of the second century B.C. the great Greek astronomer Hipparcos, relying on ancient Babylonian records was able to describe this apparent periodic motion of the celestial spheres, which, we now know, is due to the precession of the Earth axis around a line perpendicular to the zodiacal plane.

Table 1.1. Landmark achievements concerning astronomy (BC–1600).

YEAR	AUTHOR	ACHIEVEMENT
	Kepler (1571–1630)	Laws of planetary motion
	Tycho Brahe (1546–1601)	Accurate and systematic observations
1500	Copernicus (1473–1543)	"De revolutionibus orbium celestium"
	Buridan (c.1358)	Impetus theory for celestial bodies
	Alfonsine Tables (c.1252)	Improved planetary data, compiled by moorish, jewish and christian scholars under Alphonse X of Castile.
1000	Various arab scholars	Adoption of hindu numeral and decimal system, trigonometry.
	Tabit ibn-Korra (836–931)	Translator of Euclid, Apollonius, Archimedes and Ptolemy
500		
	Ptolemy (85–165)	"Almagest" Planetary data described by epicycles Use of angular parallax
0	Hipparcos (190–120)	Accurate catalog of stars Precession of equinoxes
	Aristarchus (c.150)	Distance to Earth and moon First heliocentric proposal
	Erathostenes (284–192)	Diameter of Earth
	Chih Chen (about 350)	Oldest Chinese catalog of stars
	Aristotle (384–322)	Systematization of knowledge
	Pythagoras (c.572–c.497)	Spherical Earth
-500	Thales (c.640–c.552)	Angular diameter of sun as a fraction of zodiacal circle

A full round of precession amounts to about 26000 years, i.e.

(2160 years/constellation) × (12 constellations) ≈ 26000 years   (1.2)

Some Greek philosophers fancied this very long period as the Great Year, that is, the time span after which an "eternal return" is completed and things in the universe start all over again. Of course, fairly accurate and sustained observations were required to identify and characterize this slow periodic motion in the heavens.

Very early observers could predict eclipses with fair accuracy, using their tables of lunar motion, although they did not understand their cause. Pythagoras (c.572–c.497) produced an ideal model of the universe in which this was conceived as a series of concentric imaginary spheres in which the sun, the moon and the five then known planets, as well as the Earth (and an imaginary counter-Earth) rotated around an invisible "central fire" (see Fig. 1.2), producing the "music of the spheres".

Fig. 1.2. Pythagorean model of the universe.

Erathostenes (284–192), using a very simple method, to be discussed in more detail later on, measured the central angle corresponding to a relatively long arc running South to North along the Nile river. In this way he was able to calculate within an error of only 5% the perimeter of the Earth, therefore proving its sphericity. Old reports of Egyptian sailors who had circumnavigated Africa clockwise had indicated that, in going upwards from what is now known as the cape of Good Hope, they had seen the noon-day sun shining on them from behind, contrary to what happens in the northern hemisphere. But no one then, among the learned in the Pharaoh's Court made the proper connection between this observation and the sphericity of the Earth. The estimate of the Earth's diameter by Erathostenes was crucial for the next step. Aristarchus of Samos (c.150 B.C.) gave this gigantic step forward when, making use of elegant and daring geometrical arguments, he was able to estimate the size of moon and sun relative to that of the Earth, and to make a rough evaluation of the large distance from Earth to moon, and even the normous distance from Earth to sun, in terms of Earth diameters. The relative and absolute errors were, of course, considerable, due to the lack of precise instrumentation, but his achievement can be rightly characterized[5] as "the first truly scientific step in the study of the cosmos" (more details on this later on). The large estimated size of the sun relative to the Earth led Aristarchus to make the first heliocentric conjecture recorded in history, and also to the conclusion that, if this conjecture were true, the Earth must be spinning around a certain axis of rotation.

Figure 1.3 depicts the arrangement of sun, Earth and planets according to the heliocentric proposal first intimated by Aristarchus and rediscovered by Copernicus many centuries after.

Hipparcos (190–120 B.C.), considered the greatest astronomer in antiquity, developed simple trigonometrical methods to determine angular positions of celestial bodies. He realized that astronomy requires systematic and accurate observations extended over long time spans, made use of old Babylonian observations and compared them with his own, discovering the precession of the equinoxes every 26000 years. Disregarding the heliocentric proposal of Aristarchus, he devised a geocentric scheme of cycles and epicycles (compounding of circular motions) in pursuit of the old greek aim of "saving the phenomena", i.e. providing a theory which could account for all the observed movements of sun and moon around the Earth. Ptolemy (A.C. 85–165), more than two centuries later, extended this scheme of epicycles to the planets, being able to predict their motion with considerable accuracy. His thirteen volume treatise, known by the name used by their later arab translators, the "Almagest", summarized a considerable amount of ancient astronomical knowledge, and became the most authoritative book in astronomy for many centuries. Among other achievements, Ptolemy was able to make an accurate measurement of the distance from Earth to moon by means of the technique of angular parallax (see Fig. 1.4).

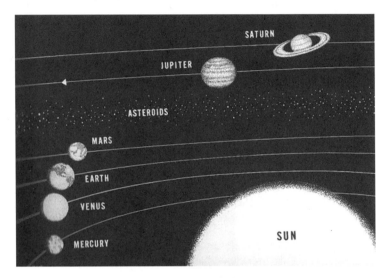

Fig. 1.3. Heliocentric System.

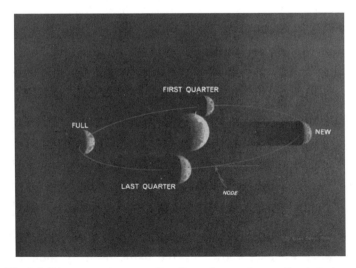

Fig. 1.4. Distance to the moon $D \approx d/\tan\alpha$ by angular parallax technique.

For a long period, after the fall and subsequent dissolution of the Roman Empire, astronomy as well as other branches of learning, in the West, became dormant. In a not too long period of time, and due probably in no small measure to generalized moral and civic decay, accompanied by the "Volkerwanderung" of the northern germanic peoples, the european population decreased steeply. Agriculture, manufacturing and commerce decreased also, and not surprisingly, so did cultural life, including preoccupations with the motion of the heavenly spheres.

By the ninth and tenth centuries of the Christian Era, arab Scholars in Baghdad and Corduva were very active in preserving, translating and commenting much of the classic Greek heritage. Tabit ibn-Korra (836–931) was the translator of Euclid, Apollonius, Archimedes and Ptolemy. The arabs made also an invaluable contribution to astronomy by adopting the hindu numeral and decimal system, which made uncumbersome the handling of very large numbers for distances and times, such as they often appear in the discussion of the heavens. In A.D. 1252, fifty moorish, jewish and christian scholars compiled and improved a complete set of planetary data in Toledo, under the patronage of Alfonse X of Castile. In the thirteenth century a new vitality was clearly discernible in medieval Europe. New beautiful cathedrals and new universities were built up. Renewed interest in all branches of learning became evident. And also in nature, as God's handiwork. We may quote the words of St. Francis of Assisi:

Blessed be Thou, my Lord, by all Thou has created and specially blessed by our brother sun which shines and opens the day, and is beautiful in its splendour; it carries on thru heavens the news of its author. And by our sister moon, and by the clear stars that your power created, so limpid, so good looking, so alive as they are. Blessed be Thou, my Lord!

A century later John Buridan (?–c.1358), chancellor of the University of Paris, almost unadvertedly, introduced a new concept in dynamics, the concept of impetus, anticipating Newton's first law, which he immediately extended to the motion of heavenly bodies, and , as it was well documented first by Pierre Duhem and more recently by Stanley L. Jaki, started a chain of events which would culminate in

Copernicus (1473–1543) "De revolutionibus orbium celestium", and later in Newton's (1642–1727) "Principia". On the authority of Aristotle, the philosopher "par excellence" among both christian and arab scholars in the late middle ages, motion in the planetary world was somehow directed by the more perfect motions in higher spheres, and so on, up to the outermost sphere of fixed stars, indistinguishable[6] from the prime mover. This implied a refined animistic and pantheistic world view, incomparably more rational than the ancient world views of Babylonians and Egyptians, among others, but a world view, nonetheless, hardly compatible with the idea of "inertial motion" which is implied in Buridan's concept of "impetus". This was therefore a momentous breaking point in the history of astronomy and cosmology in general, which was to bear fruit centuries later in the hands, first of Copernicus and then of Newton.

"De revolutionibus orbium celestium" by the polish "canonicum" Nicolaus Copernicus (1473–1543) had a truly revolutionary impact in the annals of astrophysical cosmology, which rescued from oblivium the old heliocentric proposal of Aristarchus, and did much more, incorporating the idea of inertial motion to the description of the wanderings of planetary bodies, but left for future scholars the mathematical description of these wanderings.

Tycho Brahe (1546–1601), the son of a rich Danish nobleman, and his collaborators, patiently and competently compiled, over a period of twenty years, the most accurate and complete set of astronomical observations the world had ever seen. His attitude towards Copernicus work was conservative, and he disliked the idea that the Earth was moving. There were some good reasons for this attitude, since his precise observations did not show any "parallax" for the bright, presumably close, stars against the background of the faintest and more distant stars (see Fig. 1.6). Only after almost two centuries were astronomers able to actually see the minute parallax of nearby stars due to their enormous distances from the sun. At the death of Tycho his records were passed on to Johannes Kepler (1571–1630), who had been his last assistant. Kepler spent nearly a decade trying to fit Tycho Brahe's accurate observations, articularly of Mars, into a pattern of circular heliocentric motions.

Fig. 1.5. N. Copernicus.

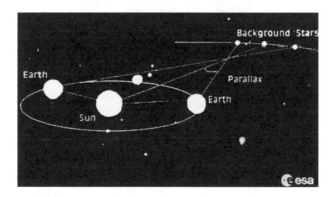

Fig. 1.6. Trigonometrical parallax of a nearby star.

A circle, as well as a sphere, were the kind of perfect geometrical objects then considered exclusively suitable to fit exactly the perfect harmony of the heavens. At last, he conceived the idea that Mars's orbit was an ellipse, with the sun at one focus of it, as developed in his greatest book, the "Astronomía Nova". This discovery led him to the three famous laws of planetary motion which bear his name. He was fully aware of the debt to Tycho's painstaking observations to achieve this. According to Kepler,[7] "The mind grasps a thing all the more correctly the closer it is to sheer quantities, that is, to its origin. The farther is a thing from quantities, the greatest is its share of darkness and error". His 3rd. law, which presupposes the other two, states that the ratio of the cube of the semimajor axis of the elliptical orbit (average distance to the sun) to the square of the period (time of one revolution) is constant for all planets, bringing unity to where there was previously a jungle of epicycles and deferents. This law was arrived at by Kepler in an empirical way and later served as the final touchstone for Newton's theory of gravitation (see Fig. 1.8).

Fig. 1.7. J. Kepler.

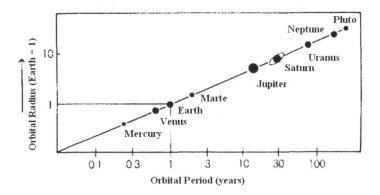

Fig. 1.8. Pictorial sketch of Kepler's laws.

1.3. From Galileo and Newton to Kirchhoff

Galileo Galilei (1564–1642), considered by many historians of science as the founder of modern physics and astronomy, made contributions certainly important in both fields, but his work, in a proper historical setting,[7] is not conceivable without the advances made by the men discussed in the previous section. He seems to have been the first man to make use of the telescope for astronomical observations, discovering the four largest satellites of Jupiter and the phases of Venus, which he used in support of the Copernican theory. This theory had been taught at several european universities, Salamanca being an outstanding example, before Galileo, but it became a subject of hot theoretical polemics between Galileo and other Italian contemporary scholars, and the incident ended up, temporarily, with an ecclesiastical injunction forbidding Galileo to teach the Copernican system.

Fig. 1.9. Galileo Galilei.

Nevertheless he viewed the book of nature as complementary with Scripture in pointing out to the existence of a Creator, and to the human mind as a most special product of the same Creator of all. As it is well known, Galileo combined the empirical and the deductive method to describe the motion of uniformly accelerated motion, reaching the conclusion that the time of free fall of bodies falling to ground from the same height is the same, regardless of their weight (contrary to the Aristotelian tenet). The law relating free fall distance (height = h) and free fall time (t),

$$h = \left(\tfrac{1}{2}\right) g t^2 = (\text{const.}) t^2 \qquad (1.3)$$

had been anticipated several decades before by the Spanish dominican friar Domingo de Soto (1494–1560), but only after being rediscovered and exploited by Galileo became widely appreciated. Galileo's work in Mechanics plowed ground for the coming fundamental work of Newton (see Table 1.2).

The great Sir Isaac Newton (1642–1726) came to be the unifier of Mechanics and Astronomy, with his famous theory of universal gravitation. In Einstein's words in the foreword to a new edition of his "Optics", Nature was to him an open book, whose letters he could read without effort' ... In one person he combined the experimenter, the theorist, the mechanic, and not least, the artist in exposition". His initial research was in Optics, being the discoverer of an admixture of distinct colours in white light after being refracted through a prism, and the first modern proponent of a corpuscular theory of light. But his most prominent achievements are to be found in Mechanics and Astronomy. In his "Principia", composed in the short period between autumn of 1684 and spring of 1686, he displayed his consummate skills in theoretical physics, of which he was pioneer and master. There he solved the problem of the Keplerian motion of the planets, and the made use of the fact that a homogeneous gravitating sphere attracts at points outside it as if all its mass were concentrated at its centre. The force of attraction was postulated by him as it is well known, as given by

$$F_g = GMm/r^2 \qquad (1.4)$$

where G is a gravitational universal constant, M the mass of the central sphere, m the mass located at the point in question, and r the distance from the centre of the sphere to the point. He tested the inverse square law of gravitation against the Moon's motion with spectacular success, employing the recently improved measure of the Earth's radius by Picard.

Table 1.2. Outstanding astronomical discovers (1600–1900).

YEAR	AUTHOR	ACHIEVEMENT
	Kirchhoff (1824–1887)	Spectroscopical analysis
	Fraunhofer (1787–1826)	of solar and star light
1900		
	Bessel (1784–1864)	First direct measurement of distance to
	Struve (1793–1864)	a star (Lirae and 61 Cygny)
	Gauss (1777–1853)	"Theoria motus corporum celestium"
		Analysis of perturbations of planetary orbits
	Olbers (1758–1840)	Paradox of dark sky at night in an infininite (?) universe.
1800	Herschel, F.J (1792–1871)	Improved large telescopes
		Resolution of Milky way in stars
	Herschel, W. (1738–1822)	Discovery of Uranus
		Discovery of binary stars: application of Kepler's laws outside solar system.
	Laplace (1749–1827)	
	Lagrange (1736–1813)	Celestial Mechanics
	Euler (1707–1783)	
1700	Lambert (1728–1777)	Recognition of Milky Way's shape
	Bradley (1693–1762)	Aberration of starlight
		Improved catalog of stars
		Nutation of Earth
	Halley (1656–1742)	Catalog of stars of Southern hemisphere
		Prediction of return of comet Halley
		Discovery of "proper motion" of stars
	Newton (1642–1727)	"Principia" (1687)
		Universal theory of gravitation
		Unification of Mechanics & Astronomy
		Theory of tides
	Galileo (1564–1642)	"Dialogue" (1632)
		Support of heliocentric theory
1600		First Use of telescope: sun spots

This was certainly a great leap forward, from the fall of a mere apple to the "falling" of the Moon, held in place by the centrifugal force due to its circular motion, and beyond.

In the "Principia", Newton rejected the un-scientific view that our primary experiences about material, impenetrable bodies, all come from pure reason, as well as the opposite extreme view that science is made up only of pure observations. He gave[9] a clear account of the moon's motion, tides, the precession of the equinoxes and the equivalence of gravitational and inertial mass without any reference to hypothetic "ad hoc" mechanisms at all, like many of his predecessors while making use of the best data available to him. He followed an epistemological "middle road" between pure rationalism and pure positivism.

Halley (1656–1742), Newton's contemporary and Bradley (1699–1762), some years afterwards, made important[10] practical observations and contributions to experimental astronomy. Among other things, Halley made an extensive catalog of stars visible in the southern hemisphere, studied the highly exocentric trajectories of comets and predicted with very good accuracy the return of the comet which was afterwards named after him. He had noted the similarity on the orbits of comets observed in the years 1456, 1531, 1607 and 1682 (at time intervals of about 75 ½ years) and correctly predicted the coming back of the same comet in 1758, which was in due time observed, sixteen years after his death. Up to this time it was generally believed that the so called fixed stars remain absolutely fixed with respect our sun, but his careful measurements of Sirius, Procyon and Arcturus convinced him that the stars have characteristic random "proper motions". Bradley, who succeeded Halley as astronomer Royal, made several important contributions. He discovered the so-called "aberration of starlight" due to the fact that the Earth's annual motion in its orbit compounds in the observing telescope with the finite velocity of light, previously discovered by the Danish astronomer Roemer (1644–1710). He found this unexpectedly while doing methodic measurements in the star γ Draconis in the hope of detecting the stellar parallax, which, as we now know required much higher precision. Another important discovery of his, the nutation of the Earth's axis, due to the changing direction of the gravitational attraction due to the Earth's bulk at the equator, arose out from his work on the aberration of light (see Fig. 1.10) Reasoning from Newton's attribution of the precession of the equinoxes to the joint gravitational action of Sun and Moon on Earth, he concluded that nutation must arise from the fact that the Moon is sometimes above and sometimes below the ecliptic plane, and it should show the periodicity of the moon's nodes (about 18.6 years).

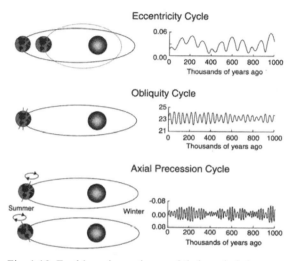

Fig. 1.10. Earth's main motions and their periods in years.

A very important step forward in the understanding of the cosmos was given in 1749, when the young John Heinrich Lambert (1728–1777) conceived the idea that the Milky Way is a lentil shape conglomerate of stars which looks the way we see it when viewed from the Earth's position, around a sun located in its main plane. This rotating and stable arrangement of stars was contradiction fee, unlike the one originally proposed by Newton for the system of stars, and became the basic pattern[12] of his description of the universe in the "Cosmological Letters" Lambert, who later became a member of the Berlin Academy of Science was a contemporary of Kant (1724–1804), who as an "amateur" in astronomy, had also conceived the Milky Way as a flattened spherical arrangement of stars. Kant, however, sought its stability in a compromise between attractive and repulsive forces, foreign to Newton's gravitational theory. Lambert was aware of the fact that an infinite universe suffered from the gravitational paradox. His work in cosmology, unlike his work in mathematics and optics, seems to have made little impact, until very recently. The realization that there are many more galaxies besides our own had to wait nearly two hundred years, until the beginning of the twentieth century.

Mention must be made, of course, of the great contributions by Euler (1707–1783), Lagrange (1736–1813) and Laplace (1749–1827) who carried forth in its fullness the application of Newton's mechanical principles to the study of the planetary system, bringing almost to perfection the science of "Celestial Mechanics".

W. Herschel (1738–1822), and his son F.J. Herschel (1792–1871) after 1816, made some of the most outstanding contributions to observational astronomy and to cosmology in post-newtonian times. W. Herschel constructed his own 5 and 1/2 feet telescope, and after becoming Court Astronomer to the King of England, his very large (20 feet) reflector telescope, with which he was able to resolve the Milky Way into individual stars. In 1787 he discovered Uranus, the next to last major planet. He also made the discovery of double star systems (binary stars),and made then an extensive catalogue of binary stars (moving around each other and therefore subject to Kepler's laws), which in later years served to determine the vast interstellar distances within our own galaxy, and, from extensive counts, the overall size and shape of the Milky Way. He also made the discovery[13] of the intrinsic motion of the Sun within our own galaxy, and was able to investigate many nebulae, globular clusters, and true galaxies ("island universes") through space, though, at the time, it was not easy to decide whether they were comparable in size and star number to our own. F.J. Herschel helped his father in reducing observational data from the Northern Hemisphere, and made a journey with his family to Cape Town to do similar investigations in the Southern Hemisphere, where he spent four years. The Herschels are remembered also for their pioneer work on astronomical colour-photometry and for the use of photographic techniques. To make clear his dislike for the purely empiricist approach to reality, W. Herschel affirmed[14]: "Half a dozen experiments made with judgment by a person who reasons well, are worth a thousand random observations of insignificant matters of fact".

Other great nineteenth century astronomers were Olbers (1758–1840), who pointed out the famous Paradox of dark sky at night in an infinite (?) universe, Gauss (1777–1853) who, in his "Theoria motus corporum celestium" made a penetrating analysis of the perturbations of planetary orbits by bodies other than the Sun, and Bessel (1784–1864), who finally solved the problem of the parallax of nearby stars by measuring the parallax (and therefore the distance) to 61 Cyngy, which by his good luck, was one of the stars nearest to our Sun.

By the end of the last century Kirchhoff (1824–1887) was making spectroscopical analysis of light from the Sun and from other star, using previous data by Fraunhofer (1787–1826), which confirmed that star surfaces were filled with the same kind of chemical elements observable here in the Earth's surroundings.

1.4. The XX Century

The XX century is witness of a vast increase in the range of both astrophysical observations and theoretical work. It marks the advancement from consideration of merely our own Milky Way, to close attention to the entire observable universe. We can consider it as the true century of Astrophysical Cosmology (see Table 1.3). During the late XIX century and early XX century, considerable information was gained on the nature and composition of stars. Even when viewed with a large telescope a star is no more than a bright point of light, except, of course the Sun. But the sun can serve to us as a guide to the characterization of other stars. Setting on the horizon, the sun looks as a compact disk (the photosphere), surrounded by a hot semitransparent atmosphere, which glows with a dull red light (the chromosphere). When the moon covers the photosphere during an eclipse, the chromosphere becomes visible as a shining ring around the sun. The atmosphere of the sun as well as the outer corona, unlike its core (made up of "plasma"), is composed of atoms, like neon, iron, etc, which can absorb sunlight, hold it for a fraction of a second, and then reemit it with a characteristic wavelength and frequency. The wavelength (or frequency) carries on with it the trademark of the excited atom. Analysis of sunlight's spectra showed that some of the emitting ions are heavily ionized, having lost, f.i., thirteen electrons, which is an indication of the fact that their equivalent temperature is of the order of several millions kelvin. This gives us an idea of the tremendous amounts of energy being continuously liberated by fusion processes in the sun's interior, and similarly, in other stars. It is well known that the gravitational pull of the solar matter towards its centre is checked by the radiation and gas pressure originated in the fusion processes. We can learn a lot about the characteristics of a star by looking at the radiation emitted by it, which contains a continuous part indicative of the star's temperature and a discontinuous set of lines, which shows the trademarks of the atoms present in its atmosphere. A preliminary classification of stars by colour (or, equivalently, surface temperature) was designed by the first decade of the twentieth century.

Eight spectral types were arbitrarily defined and labeled by the capital letters OBAFGKMN (mnemotechnic rule: look for the initial letters of words in the sentence "Oh! be a fine girl, kiss me now") The surface temperature, the colour and one typical example are given in Table 1.4.

Shortly afterwards it was realized that the temperature of a star, which depends upon its bulk, is related to its intrinsic luminosity, i.e. the luminosity, or amount of light radiated per unit time which would be observed when the star is removed from its actual distance and placed at a standard distance from the observer, the same for all stars.

The Herzsprung-Russell diagram, which leads to a relationship between luminosity and temperature, was discovered in 1911–1913 (see Fig. 1.11).

In this diagram the "main sequence" corresponds to the hydrogen burning phase in the life of a star. Very heavy stars burn hydrogen quickly and then, after undergoing violent explosions (supernovae) in which a faint, distant star becomes temporarily extremely bright, and afterwards leaves the main sequence. Very light stars, on the other hand, burn hydrogen slowly. Our sun occupies and intermediate position on the main sequence and is now approximately half way in its hydrogen burning phase, which for stars with a mass alike the sun's mass is of the order of 10^{10} years.

Table 1.3. Cosmological advancements in the XX century.

Year	Event	Description
	(1987)	Supernova 87
1980	(1980's)	Connection between Cosmology and Elementary Particle Physics.
	Apollo XI Mission (1969)	First manned landing on the moon.
	Bell (1967)	
	Wagoner, Fowler, Hoyle (1967)	First "pulsar" discovered Theory of light clements primordial nucleosynthesis.
	Penzias–Wilson (1965)	Observation of 3 K background radiation
1950	Hazard (1952)	First "quasar" discovered
	Burbridge, Fowler Hoyle (1957)	Theory of nucleosynthesis of heavier elements in the core of stars.
1940	Alpher–Herman–Gamaw (1948)	Prediction of remanent background radiation Primitive theory of cosmic nucleosynthesis.
1930	Hubble–Humason (1928–30)	Systematic measurements of galaxy distances & recession velocities
	Lamaître (1927)	Primitive "Big Bang" theory General solutions of Einstein's eqs. ($\Lambda \neq 0$)
	Friedmann (1922)	General solutions of Einstein's eqs. ($\Lambda \neq 0$)
1920	Eddington (1919)	Organized eclipse expedition which tested theory of General Relativity.
	Einstein (1917)/ De Sitter (1917)	General relativistic cosmological equations/ Special solution to Einstein's equations predicting universal expansion.
	Slipher (1914)	First report of observation of a receding galaxy (Andromeda)
1900	Hertzsprung–Russell (1911–13)	Diagram displaying relationship between luminosity and temperature (color) of typical stars.

Table 1.4. Classification of stars by spectral type.

Spectral Type	Pothosphere temperature	Color	Example
O	55,000	Blue	Orion's sword
B	25,000	Pale blue	Spica
A	11,000	Blue-white	Sirius, Vega
F	7,000	White	Procyon
G	6,000	Yellow	SUN, Capella
K	5,000	Pale orange	Aldebaran
M	3,300	Orange	Antares
N	2,800	Red	None visible

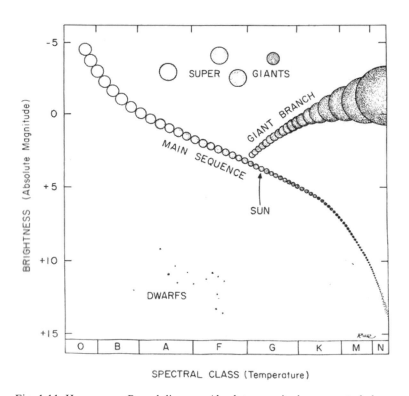

Fig. 1.11. Herzsprung–Russel diagram: Absolute magnitude vs. spectral class.

In 1914 Slipher,[5] an American astronomer which was investigating the Andromeda nebula, as a good possible candidate for a new solar system in formation, discovered that the spectral lines appearing in the light coming from this nebula were strongly shifted with respect to the corresponding lines of an emitting body at rest with respect to the observer. He correctly concluded that the Andromeda system, very far away from our system of stars, was receding at a very high speed away from us, interpreting the redshifting of spectral lines as due to the Doppler effect. This was immediately recognized as an epoch making discovery by the members of the American Astronomical Society attending the meeting in Evanston, Illinois at which Slipher reported his finding. Edwin Hubble, then an astronomy student attending the meeting, took good note of the significance of this observation, and conceived the idea (to be carried on to effect more that a decade later) of making a full scale exploration of similar systems through the vast spaces of the observable universe.

At about the same time, in the year 1917, Einstein, completely unaware of Slipher's observation was busy trying to apply his newly developed theory of General Relativity to the description of the whole universe. He was able to arrive at general relativistic cosmological equations, in which he introduced, somewhat arbitrarily, a repulsive "cosmological" term, which could compensate the universal attraction and therefore make room for a static distribution of star systems. Almost simultaneously, De Sitter, which had received a copy of Einstein's paper, obtained a special solution of the general relativistic equations which predicted a universal expansion for the cosmos as a whole.

Just after the end of the European War 1914–1918, which had forced a postponement of the initial plans, Sir Arthur Eddington, a leading british astronomer and theoretical physicist, organized an eclipse expedition to verify the bending of light by gravity, as predicted by Einstein's General Relativity. This occurred in 1919. The bending of light was verified, and Einstein became overnight a world celebrity. At the same time this confirmation of General Relativity provided strong indirect support for the favourable expectations concerning its application to the description of the cosmos.

In 1922, Alexander Friedmann a Russian meteorologist and mathematician, and, independently, five years later, George Lemaitre, a Belgian priest and mathematician, who would become in the future President of the "Ponticia Academia Scientianum" in Rome, arrived at a general set of solutions to Einstein's dynamical equations for the cosmos, which included three cases: closed (oscillating) universe, with space curvature $k > 0$; euclidean (expanding just at escape velocity) universe, with $k = 0$; and open (expanding at higher velocity) universe, with $k < 0$. Also, Lemaitre was the first to describe the initial stages of the dense, exploding universe as a primeval atom, which can be considered as the primitive version of the "Big Bang" theory.

The years 1928–1930 witnessed the successful systematic measurements[16] of galaxy distances and recession velocities by Hubble and Humarson, in the 100 inch mirror Mount Wilson telescope, which lead to the discovery of Hubble's law (see Fig. 1.12). These investigations were carried on to further distances with the new 200 inch mirror telescope at Mount Palomar. The original measure of Hubble's constant,

$$H_0 = \dot{R}/R = (v/d) = 500 \text{ Km/ sec /Mpc} \cong [2 \times 10^9 \text{ years}]^{-1} \quad (1.5)$$

where one Mpc (megaparsec) = 3.26×10^6 light years, was successively corrected for errors in the determination of the distance scale, leading to its presently accepted value,

$$H_0 = (75.3 \pm 30) \text{ Km/sec /Mpc} \cong [13 \pm 4 \times 10^9 \text{ years}]^{-1} \quad (1.6)$$

which implies an increase in size of the order of seven times the original estimates. Large errors like this are not uncommon in the history of physics, but what was epoch making was the formulation of an important cosmic law, Hubble's law. After the pioneering numerical estimates were performed, more refined measurements would come in due time.

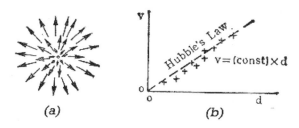

Fig. 1.12. (a) Expansion of galaxies at radial velocities increasing with distance (b) Schematic representation of *v* (velocity) vs. *d* (distance) observed by Hubble and Humarson.

During the period preceding the Second World War, and in wartime, the attention of most physicists, both in Europe and America, was primarily focused in research topics directly or indirectly related to the war effort, such as the development of radar systems and the "Manhattan" project to build a fission bomb.

Two young collaborators of George Gamow, who were investigating in 1948 the possibility of nucleosynthesis in the early stages of the expanding universe, predicted that a remnant cosmic radiation background of about 5 K should be present in the universe as a result of the cooling down of the very intense (high temperature) radiation background at the time of nucleosynthesis. Alpher's dissertation was published the same month as the joint paper in the British journal "Nature" in which the 5 K prediction was reported. At the time, no one seems to have paid much attention to this bold prediction, which seventeen years later was fully vindicated by the experimental finding of the radioastronomers A.A. Penzias and R. Wilson, who, as is often the case, were looking for another thing.

In 1957, Burbridge, Fowler and Hoyle worked out a theory of the nucleosynthesis of the heavier elements in the core of stars, which was able to account satisfactorily for the main data available from spectroscopic analyses. A second or third generation star, like our sun, contains from the beginning a small percent of heavy elements resulting from supernova explosions of stars from a previous generation.

Hazard (1962) discovered the first quasi-stellar object ("quasar"), characterized by its extraordinary brightness, much larger than that of an ordinary galaxy, and its large recession velocity. After this, many other "quasars" were discovered by various investigators, giving clues, not yet completely understood, of the state of the cosmos at early times.

The detection of an isotropic and uniform 3 K blackbody radiation coming from all directions of the universe, experimentally performed by Penzias and Wilson in 1965, confirmed the prediction of Alpher, Herman and Gamow, made on the basis of the Big Bang theory some years before, and was instrumental in favouring the Big Bang theory over its main rival theory at the time, the so called steady-state theory. It is worth noting that the steady-state theory assumed continuous creation out of nothing in the universe, thus violating the principle of matter/energy conservation. This is a clear indication of how far some, otherwise competent theorists were willing to go in their theoretical efforts to describe the cosmos.

In 1967 Bell discovered and characterized the first "pulsar", extrange pulsating radio source which was later identified with a rotating neutron star, and was located relatively close to us in our own galaxy. These "pulsars" can be the result of collapse after a supernova explosion of a star with appropriate initial mass. In the same year, Wagoner, Fowler and Hoyle produced the theory of "primordial" (that is produced at very early times) nucleosynthesis of He and traces other light elements, which, unlike the heavier elements, seem to be more or less uniformly distributed throughout cosmic space.

The year 1969 witnessed, as all those with sufficient age may remember, a man on the Moon (Apollo XI Mission). The 1970's saw the awakening of theoretical efforts to connect Cosmology and Elementary Particle Physics. This provided interesting insights into the allowable number, type and degeneration of neutrinos, and into the number and properties of weekly interacting particles.

Very recently (1987) contemporary astronomers had the unique opportunity of investigating closely the initial stages of a supernova explosion outside our galaxy, the Supernova 87, with all the sophisticated instrumentation at their disposal.

Current investigations on "black holes", the so called "missing mass" (a misnomer, because we do not know whether this mass really exists), the astonishingly large photon to baryon ratio in the cosmos, etc., leave many open questions.

The story of observational and theoretic investigations on the universe, which begins to look more like a cozy place for Earth and men than the vast punctuated nothingness previously envisioned by some, goes on.

Bibliography

1. A. Einstein, "Lettres a Maurice Solovine" (Paris: Gauthier-Villars, 1956).
2   J.L.E. Dreyer, "The History of Astronomy" (New York: Dover, 1953).

3. S.L. Jaki, in "Física y Religión en Perspectiva", p. 21 (Madrid: Rialp, 1991).

4. G.S. Hawkings, "Splendor in the Sky" (New York: Harper and Row, 1969).

5. S.L. Jaki, "God and the Cosmologists" (Washington: Gateway, 1989).

6. S.L. Jaki, in "Física y Religión en Perspectiva", p. 28 (Madrid: Rialp, 1991).

7. P. Duhem, "Le system du monde" (Paris: A. Herman et Fils, 1913).

8. S.L. Jaki, "The Road of Science and the Ways to God", p. 47 (Chicago: The University of Chicago Press, 1978).

9. "A Biographical Dictionary of Scientist", p. 94 (London: A. and C. Black, 1969).

10. Ibid., p. 74 and 238.

11. S.L. Jaki, "The Milky Way: An Elusive Road for Science" (New York: Science History Publications, 1972).

12. S.L. Jaki, "The Savoir of Science" (Washington: Gateway, 1988).

13. "A Biographical Dictionary of Scientist", p. 254 (London: A. and C. Black, 1969).

14. "The Scientific Papers of Sir William Herschel" (Ed. J.L. Dreyer, London: Royal Society, 1912).

15. S. Weinberg, "Gravitation and Cosmology", p. 417 (New York: Wiley, 1972).

16. Ibid., p. 445–46.

17. R. Alpher and R. Herman, "Physics Today", Part I, p. 24 (August 1988).

Chapter 2

The Importance of Precision

2.1. The Last Word in Physics

Probably there is no such a thing as a last word in Physics. But, at a given time, the last word in the description of a certain physical phenomenon is not affordable by theory. Experiment, of course sufficiently careful, accurate and repeatable experiment, has always the last word when in conflict with theory, although experiments are never fully detached from theoretical considerations. This is a characteristic feature of Astrophysical Cosmology, as well as of down to Earth Physics and Elementary Particle Physics, which is not meant to deny that theoretical concepts are equally essential ingredients. Bernard Shaw,[1] commenting on the impact of the theory of relativity, said: "Einstein has not challenged the facts of science but the axioms of science, and science has surrendered to the challenge". Had the facts of physics challenged the axioms of the theory of the relativity, things would have gone the other way around. As a matter of fact Planck's quantum theory was validated by precise measurements of the black body radiation made by Wienn and others, Einstein's special relativity by the crucial and very precise Michelson–Morley experiments, and the General Theory of Relativity, among other things, by the good agreement of its prediction of the advance per century of the perihelion (point nearest the Sun in its orbit) of Mercury[2] with the experimental observation, about 574 arc seconds, which was only 43 arc seconds larger than the value predicted by newtonian theory.

To measure a physical quantity the observer makes a determination of its magnitude by comparison with a standard for this quantity. The choice of instrument, measuring procedure and standard unit is important to make precise measurements. To measure large distances a microscope is useless, while to measure small distances one cannot employ a telescope. The basic standard unit is usually divided in smaller and smaller subunits, (1 microunit = 10^{-6} basic units) or multiplied in increasingly large multiples (1 megaunit = 10^6 basic units). When changing the scale of measurement, care must be taken not to introduce errors. The scale of observable lengths goes over from 10^{-13} cm, the size of a proton, to 10^{28} cm, roughly the distance to the most faraway galaxies, i.e. 41 orders of magnitude. The scale of times, from 10^{-15} sec., the lifetime of some electronically detectable atomic processes, to 10^{39} sec., the estimated lower limit of the proton's lifetime, i.e. 54 orders of magnitude. The scale of masses from 10^{-30} grams, the mass of an electron, to 10^{53} grams, roughly, the mass in the observable universe, i.e. 83 orders of magnitude. Very large changes in scale can be observed or deduced in other physical quantities, f.i. from picovolts to gigavolts, 21 orders of magnitude, or from tens of nanokelvin to megakelvin, 14 orders of magnitude. Some of the experimental devices used in modern laboratories, like the Mossbauer spectrometers can have a precision sufficient to distinguish one in 10^{12} parts. Other are, of course, less precise.

More than one century ago Maxwell[3] described a prevalent mood among his colleagues in the following words: "The opinion seems to have got abroad that in a few years all the great physical constants will have been approximately estimated, and that the only occupation which will then be left for men of science will be to carry on these measurements to another place of decimals". A younger colleague[4] added: "In many cases the student was led to believe that the main facts of nature were all known, that the chances of any great discovery being made by experiment were vanishingly small, and that therefore the experimentalist's work consisted in deciding between rival theories, or in finding some small residual effect, which might add a more or less important detail to the theory". Maxwell did not go along with this view, and pointed out that more precise measurements should lead to new regions, new fields of research, new scientific ideas. In fact, within a few decades, more precise and new experimental findings opened the way to quantum theory, to special relativity, and, unexpectedly, a new experimental discovery, that of natural radioactivity, was the key to the world of nuclear physics and beyond.

The quest for qualitative precision usually requires sustained and painstaking efforts, but the rewards can be very high.

2.2. Precise Astronomical Observations

With some notable exceptions, the Greeks did not show much of a consistent appreciation for quantitative precision. And some of their major discoveries are precisely connected with the outstanding exceptions. A case in point is that of Hipparcos, the discoverer of the precession of the equinoxes. The figure given by him was $1°23'20''$/century, only $10''$ per century less than the accepted value today. What Hipparcos wanted to get was a better calendar, and to this end he was comparing the lengths of the tropical and the sidereal years. He used the data continuously recorded by Babylonian astronomers since 747 B.C. and compared them with his own careful observations. He was able to see that the Babylonian records were internally consistent and also consistent with his own records, and rightly concluded that the equinoctial point moves slowly forwards with respect to the fixed stars in the zodiacal belt, completing a full revolution every 26000 years, as mentioned in the previous section.

It may be recalled that the authority of Aristotle, who did not class physics with the exact sciences, loomed large in the medieval Christian and Islamic cultures. And he said,[5] "it is the mark of an educated man to look for precision in each class of things just so far as the nature of the subject admits". This is not an unreasonable statement in itself, but, as it was meant by Aristotle, and probably understood by his medieval readers, it implied that quantitative investigation of motion below the realm of the fixed stars, the Empirical Heavens, should content itself with a moderate precision, not too much.

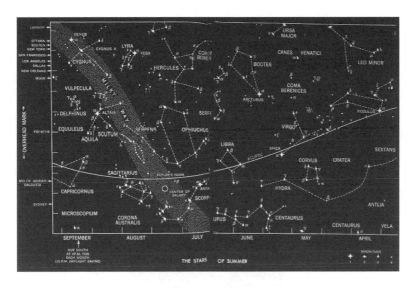

Fig. 2.1. Starts of Summer.

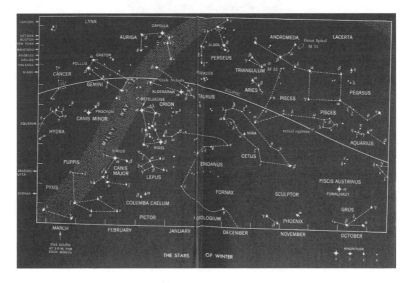

Fig. 2.2. Starts of Winter.

The best planetary positional observations of Copernicus, with errors of the order 10' (ten arc minutes) were not as good as some of the best ancient positional measurements, which attained errors of about 2' to 3' (two to three arc minutes). The predictive capabilities[6] of his system were not better than those of the Ptolemaic system. The superiority of the Copernican over the Ptolemaic system rested more on its superior simplicity than in the precision it afforded. It was therefore necessary to wait until much better observational records were made available by Tycho Brahe in order that Kepler could make decisive progress in the understanding of planetary motions.

Fig. 2.3. Tycho Brahe.

When Tycho was a fourteen years old student at the University of Copenhagen, he was impressed by an eclipse of the sun. Three years later, already in possession of the best astronomical tables available in his time, he observed one August night of 1573 that Jupiter and Saturn were so close to each other in the Sky that they had become almost undistinguishable. The Alphonsine and Copernican tables predicted this event with errors of one month and several days, respectively, which induced Tycho to commit himself to a life long effort in pursuing higher precision. He decided that better instruments, more refined methods of observation, and close attention to the errors involved, were absolutely necessary. He constructed a huge sextant, well equipped with tables to estimate the observational errors, which, among other things served to establish the superlunary position of the nova of 1572 and the comet of 1577. His long list of carefully taken data[7] set the record straight for the new coming students of the heavens, and, because of his influence, the role of the precise positions of the planets became a decisive test on the validity of the Copernican hypothesis.

After his arrival at Tycho's residence in Benastek, Kepler realized what a rich mine of gold for astronomy were the precise and carefully taken data of the Danish nobleman. "Tycho possesses the best observations and consequently, as it were, the material for the erection of a new structure", said Kepler.

Kepler's principal ambition was to be the architect of this new structure, the heliocentric system, and he knew that he had what Tycho lacked, the mathematical and physical intuition to go "exceedingly deep" under the dazzlingly diverse observed phenomena. The young Kepler could cheat with the observational data of the Copernican and Ptolemaic tables in his "Misterium Cosmographicum". But he could not do the same with Tycho's data. He could not neglect as an error a mere 8′ (eight arc minutes) because, as he said, divine mercy "has given us Tycho, on observer so faithful that he could not possibly have made this error of eight minutes" which pointed the road to "the reformation of the whole of astronomy".

In eloquent words[8] of the contemporary historian of science Staley L. Jaki: "Truly, never before had eight minutes constituted, as Kepler put it, the burden of a major astronomical work, and never before or since had the future of science depended so much on an unconditional reverence for the data of observation. That precision could come into its own in physics and assume its role of supreme arbiter over theories depended also, of course, on the conviction, growing stronger since Copernicus, that the geometry involved in the physical world must be simple. The concrete form of that geometrical simplicity was henceforth to be extracted not from one's preference for one or another geometrical figure but from the data of precision. In Kepler's situation, the hard fact of eight minutes of arc forced him to discover the true form of planetary geometry, the elliptical orbits".

The first large-scale fusion in the history of physical science of theory, bold generalization, and critically weighed experimental data was contained in Newton's Principia.

It is well known that Newton carried out with outmost accuracy his own experiments. In the first paragraph of his masterpiece he begins with a definition of the constancy of mass which, in his words, is based on "very accurately made" measurements. Mentions of "more accurate" measurements occur several times in the prefaces which he wrote for successive editions of his "Principia". This preoccupation with accuracy comes out again in connection with particular conclusions resulting from his theory. For instance, he concluded that if the Earth has a more or less uniform density, due to the continuous revolution about its axis, it must be flattened at the poles, or, conversely, must show a definitive bulge at the equator. His calculation of the resulting ellipticity (deviation from exact circularity in the meridians) gave 1/230. This is not very far off from the modern accepted value, 1/297. The polar flattening of the Earth should give rise, as Newton noted, to a series of observable phenomena, e.g. the change in the period of pendulums, the decrease of weights near he equator and the precession of the equinoxes. Pendulum experiments to detect the effects of this polar flattening were carried out in the last decades of the seventeenth century, and Newton closely scrutinized them, rejecting as "gross" those of Couplet, and praising for their "diligence and care" those of Ritcher.[9] Newton's finding of the equality of gravitational and inertial masses was supported, in his words, "by experiments made with the greatest accuracy", and it has been confirmed by twentieth century experiments to an accuracy of one part in 10^{10}. In one very particular case, the advance made by the perihelion of Mercury (at Newton's time estimated as $4'16'$ per century, certainly an small amount), Newton was wrong. He neglected it as something "inconsiderable". Later theoretical developments pointed to the perturbing effects of the larger planets, Jupiter and Saturn, as the main cause for the advancement of Mercury's perihelion, as well as that of other planets. But this was not the whole story: more than two hundred years after, it was precisely a small disagreement between the observed and the calculated advancement per century in Mercury's perihelion that provided one of the most spectacular tests of Einstein's General Theory of Relativity.

In 1845 it was found by Leverrier that the value of the advance per century of Mercury's perihelion was 574 arc seconds, i.e. almost twice the value estimated by Newton's time. Out of these 574″ there were 42″ which could not be accounted for by any realistic pattern of perturbation by the known major planets. Lescarbault, an amateur astronomer, reported having observed a small round object across the Sun, which was interpreted by Leverrier as a new planet which should have been at about half the distance of Mercury from the Sun. This tentative interpretation, on Leverrier's part, which was never confirmed by further systematic observations, is understandable taking into account that he had discovered previously a true new planet, Neptune, whose orbit was beyond that of Saturn. In 1884 the amount of unaccounted for advance of Mercury's perihelion was determined more exactly by Newcomb as being 43″. Several wild proposals were made by various astronomers to justify the discrepancy, but the matter remained obscure.

A consistent solution came up only in our century, when Einstein showed that Newton's gravitational law, till then the most universal and revered physical law, could be considered valid only as a limiting case (the case of sufficiently long distances away from the central mass) of the more general law provided by Einstein's theory of General Relativity. Einstein was certainly justified when he wrote[10] to Sommerfeld in 1915:

"The marvellous thing that I experienced was the fact that not only did Newton's theory result as a first approximation but also the perihelion motion of Mercury (43″ per century) as second approximation" (See Fig. 2.4).

However, as noted by S.L. Jaki,[11] "the more momentous is the success of a theory, the more inexorable is the scrutiny to which it is subject". Recent estimates point out that as much as ten per cent of the value of the advance in Mercury's perihelion might be due to the flattening of the Sun's inner core, hidden under the bright gas envelope. Thus, the 43″ per century result of General Relativity cannot, as yet, be taken as its definitive proof. It would not be surprising that the systematic observations carried out during the total solar eclipse of July 11, 1991, which was closely observed from one of the world's largest concentration of high powered telescopes at Hawaii's Mauna Kea, were to throw some light on the problem of the Sun's hidden flattening, and therefore on its possible contribution to the perihelion motion of Mercury.

It is worth recalling that to settle the classic proof (detection of the stellar parallax) of the heliocentric theory, astronomers had to fight very hard to improve the angular resolution of telescopes. The final success was due to the sustained efforts of several generations.[13] The first serious attempt was made by Hooke in 1669, but the objective lens of his telescope fell from the roof and shattered.

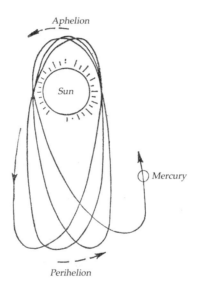

Fig. 2.4. Motion of the perihelion of Mercury around the Sun. According to Einstein's theory of General Relativity the Keplerian ellipsis does not close exactly in the gravitational field of a point mass, because the evolution of time occurs differently in the aphelion and in the perihelion. This produces an orbit in the form of a "rosetta". The orbit due to a flattened pseudosphere at the point of the Sun would produce a very similar orbit.

Roemer spent more than three years (1701–04) trying to measure the parallax of Sirius and Vega. He believed, incorrectly, because his experimental accuracy was not good enough, that he had detected a small annual change in the angle between both stars. Flamsteed reduced considerably the error in measurements of angular separation between stars and thought that he had observed a parallax of 40″, but, almost surely, what he had seen was the aberration of light, later discovered and characterized by another British Astronomer Royal, Bradley. Roemer's data were reexamined in 1727 by P. Horrebow, who published his supposed finding of the stellar parallax in a paper entitled "Copernicus triunfans". At about the same time Bradley reported that the yearly variation in the direction of stars was approximately 40″, and he attributed it, as mentioned previously, to the annual change of the relative direction of the Earth's velocity with respect to the direction of the velocity of the light coming from a given star, which was therefore implicitly recognized as a finite velocity. Bessel and Struve had at their disposal the best instruments of their time (mid nineteenth century) for determination of the star parallax, if there was any. Both, the Konisberg heliometer and the Dorpat refractor, were due to the skilful craftmanship of Fraunhofer, then the leading expert in optical instruments. By good luck they directed their instruments to two nearby stars (not necessarily the brightest) and they detected, respectively, 12/100 of an arc second (for α Lyrae, observed by Struve in 1837) and 29/100 of an arc second (for 61 Cyngy, examined by Bessel in 1838). In this way it came to an end the longest and most persistent effort in the history of astronomy to measure with precision a crucial magnitude. Copernicus' vision and Newton's theory had been fully vindicated by precise and difficult observations.

Another very import observation,[14] connecting the Earth and the Sun's chemical composition, was due to very precise measurements made by W.A. Miller on spectroscopic absorbtion and emission lines. He was able to establish conclusively the coincidence between the double dark D line in the solar spectrum and the double bright line emitted by sodium on Earth at exactly the same wavelength, using such an extended spectrum that the doublets were seen widely separated for the outmost precision.

When questioned by Lord Kelvin about the reliability of Miller's finding, Stokes, after having seen the experimental data, replied: "I believed there was vapour of sodium in the sun's atmosphere".

Michelson's case[15] is an outstanding example of achievement in precision.

Apart from his leading role in the performance of the famous Michelson-Morley experiment, showing the apparent constancy of the velocity of light regardless of the Earth's motion with respect to the hypothetical ether, Michelson gave abundant proof of his scientific genius with other truly astonishing achievements in precision. He determined the length of the standard meter with an error of one part in 10^7 in terms of the wavelength of cadmium. With his larger diffraction grating (9.4 inch), containing 117,000 lines, he was able to get the highest precision ever achieved. He made measurements of the speed of light reducing the probable error to 1/50 that of previous measurements. He was able for the first time to measure diameters of large stars, some of them later called red giants, capable of encompassing within their boundaries a large fraction of our solar system in spite of the fact the their mass was not very much higher than the Sun's mass. He even was able to verify, in some particular instances, the tidal effects produced on the Earth's crust which Newton considered beyond the range of observational possibility. But his name will be remembered by history for his demonstration that no such a thing as an "ether's drag" had an effect on the propagation velocity of light. Maxwell, who had showed that the effect, depending on the ratio between the Earth's orbital velocity and the velocity of light, and corresponding therefore to a change of one part in 10^8, considered it "too small to be observable". But it was this type of challenge that appealed to Michelson. As it is well known, his superb interferometer, capable of detecting a small fraction of the estimated change, gave a null result (see Fig. 2.5). More than half a century later, renewed versions of the Michelson–Morley experiment using masers and lasers carried on forward the accuracy of the experiment to being able to detect changes of one part in 10^{12}, an unsurpassed feat in precision. In 1931 Albert Einstein addressed Michelson at a festive gathering in the following words[15]: "Without your work this theory would today be scarcely more than an interesting speculation; it was your verification which first set the theory on a real basis". Incidentally, the Nobel Prize was awarded to Michelson for his work on the calibration of the standard meter in terms of cadmium wavelengths, not for his now much better known experiment, in

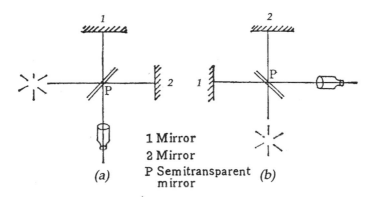

Fig. 2.5. Michelson–Morley experiment. Light reflected from mirrors 1 and 2 produces observable interference fringes. The velocity of light should compound differently with the velocity of the Earth in cases (a) and (b) rotated by 90° from (a), if there were such a thing as an ether drag, but no effect was observed neither in summer nor in winter in spite of the fact that the instrument's accuracy was more than sufficient to detect changes ten times smaller than those to be expected on the basic of the known translational velocity of the Earth (about 3×16^6 cm/sec).

collaboration with Morley, confirming the constancy of the speed of light.

To do justice to all precise measurements of physical constants of cosmological importance carried out since the end of last century would probably require a large size encyclopedia. Precision and care lead Hess to discover the penetrating flow[16] of cosmic rays bombarding our upper atmosphere from outer space, some of them with energies as high as 10^{20} eV (much higher than those obtained in the largest particle accelerators), and, more recently, Penzias and Wilson to discover the microwave radiation background, at present the most accurately known pointer to an early Big Bang. These are only two among many other recent, and not so recent, fruitful adventures in precision.

It can be noted that in most cases the uncertainties are less than one part per million (ppm), the notable exceptions being Newton's gravitational constant (G), by far the less precisely known, and Boltzmann's constant (k_B). Digits in parenthesis are one -standard-deviation- uncertainties in the last digit, obtained from least squares adjustments of more than 200 different measurements.

Table 2.1. Gives, by way of example, the 1986 uncertainties[17] of selected universal constants as recommended by CODATA (Committee on Data for Science and Technology of the international Council of Scientific Unions).

Quantity	Symbol	Value	Units	Relative Uncertainty
Speed of light in vacuo	c	299792458	m sec^{-1}	exact
Newton grav. constant	G	6.617259(85)	10^{-11} m^3 kg^{-1} sec^2	128
Planck constant	h	6.6270755(40)	10^{-34} J scc	0.60
Electron charge	e	1.30217733(49)	10^{-19} C	0.30
Electron mass	m_e	9.1093897(54)	10^{-31} kg	0.59
Proton mass	m_p	1.6726231(10)	10^{-27} kg	0.59
Avogadro constant	N_A	6.0221367(36)	10^{23} mol^{-1}	0.59
Boltzmann constant	k_B	1.380658(12)	10^{-23} J K	8.5

2.3. The New Generation of Telescopes

Since Galileo's time an improvement by a factor of 10^9 has been achieved in sensitivity of telescopic astrophysical observations. However, angular resolution, limited by atmospheric turbulence, has increased only by a modest factor of about 10^2 under optimum conditions, until very recently. But new optical techniques, such as the use of adaptive optics and interferometry are producing spectacular improvements which may give rise to a further increase of 10^2 in angular resolution for optical and infrared telescopes, approaching the inescapable diffraction limit given by

$$(\theta)_{\min} \approx 1.22(\lambda/D) \tag{2.1}$$

where a is the wavelength of the observed radiation, and D the largest dimension of the telescope, or the largest separation between multiple telescopes. Where previously only the wavelengths of visible light were used, now a wide range of radiation from all kinds of astronomical objects are being observed, including ranges in the radio and microwaves ($\lambda > 10^{-3}$ m), infrared (10^{-3} m $> \lambda > 10^{-6}$ m) visible (7.8×10^{-7} m $> \lambda > 3.9 \times 10^{-7}$ m), ultraviolet (4×10^{-7} m $> \lambda > 10^{-9}$ m), X-rays (10^{-9} m $> \lambda > 10^{-11}$ m) and γ-rays ($\lambda < 10^{-11}$ m) regions of the electromagnetic spectrum. The observations are made with underground, ground based, airborne and orbiting telescopes. Of course, underground and ground based telescopes can be larger, heavier and more conveniently manhandled, while airborne and spatially orbiting telescopes, of which the recently lunched HTS (Hubble Space Telescope) is the most famous, have the advantage of avoiding absorbtion and disturbances from the Earth's atmosphere.

Adaptive optics, one of the two new techniques mentioned before, consists in the real-time adjustment of optical surfaces to compensate by wavefront distortions due to atmospheric turbulence. This technique monitors the wavefront errors (with reference to a nearby bright star) in individual, appropriately sized, patches of the telescope aperture, during the short "coherence time" in which the phase of the incoming light is well defined, and then corrects them by warping the mirror. So the whole

Interferometry, on the other hand, consists in the phased operation of multiple telescopes separated by sufficiently large distances. Some important recent results using this technique include, according to C.A. Beichman and S. Ridgeway, wide-angle astrometry, milliarcsecond measurements of the diameter of stellar photospheres, a direct determination of the distance to the Hyades Cluster (one of the most important rungs in the cosmic distance ladder) and the resolving of close binary stars. Figure 2.6 gives a spectacular example of binary star resolution obtained with non-redundant warping of the 5 m. Hale telescope by means of the interferometric technique.

During the 1980's many important discoveries have been made using underground, ground based, airborne and space telescopes. The following are some prominent examples. Quasars at extremely large distances, with redshifts up to $Z = 4.73$, implying a recession speed nearly 95 percent that of light, have been observed. The action of galaxies as gravitational lenses for light coming from more distant quasars along the same path, an effect predicted by General Relativity, has been investigated. X-ray telescopes have identified possible giant black holes at the centre of some galaxies. Orbiting infrared telescopes have been able to observe disks of solid matter surrounding nearby stars as well as galaxies with one hundred times more energy radiated in the infrared than in the visible region of the electromagnetic spectrum. Early observation of the Supernova 87 provided the opportunity to detect incoming antineutrinos produced in the subsequent explosion by means of underwater Cerenkov detectors.

Fig. 2.6. Long exposure image of the Sigma Herculis system obtained by long exposure without (a) and with (b) interferometric techniques. Angular size of blurred disk in (a) is ≈2 arcsecond. Angular separation of binary stars in (b) is ≈0.070 arcseconds.

Spinning neutron stars with very regular spinning rates, as high as 1000 revolutions per second, were discovered by means of radiotelescopic observations. This kind of systems may serve very well as astronomical clocks to study cluster dynamics. Techniques analogous to the seismologic techniques used to investigate the inner Earth's regions have been used to probe the solar interior, gaining information on the extent of the solar convective zone and on the dependence on depth of the rotation speed of solar matter. The mass and the radius of Pluto, the Sun's minor satellite were determined by means of observations of Pluto's satellite Charon.

Prospects of further improvements in precise astronomical observations in the decade closing our century are indeed very high. The first 10 m telescope, the Keck telescope built in Hawaii, is expected to come into operation very soon. This will belong to the next generation of large optical and infrared telescopes constructed fourty years after the installation of the 5 m large Hale telescope at Mount Palomar in California, which was so useful in providing information on the universal recession of galaxies by E. Hubble and those workers coming after him. Improvements in infrared telescopes will include the projected coming into operation of SIRTF (Space Infrared Telescope Facility), IRO (Infrared Optimized 8 m telescope built in Mount Mauna Kea, Hawaii), and SOFIA (Stratospheric Observatory for Far Infra-Red Astronomy) which will make use of a 2.5 m telescope system mounted in a modified Boeing 747 aircraft flying in the stratosphere above 99 percent of the atmosphere's water vapor, which acts as an strong absorber of infrared radiation.

Infrared astronomy has benefited greatly from recent major technological advances, including the development of infrared radiation sensitive solid-state detector chips, and methods developed for the long-duration containment of superfluid liquid helium in space, which result in improved sensitivity by cryogenic cooling. Improved infrared telescopes are capable of providing valuable information on cool states of matter (most common stars are cooler than the Sun and emit a large amount of energy in the infrared), and on hidden regions of the universe blocked by cosmic dust in the visible region of the spectrum (like the center region of our own galaxy, as shown in Fig. 2.7).

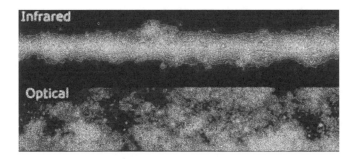

Fig. 2.7. View of the galactic plane as it appears in the far infrared (a) and optical (b). The interstellar dust obscures almost completely the gallactic plane at optical wavelengths. On the other hand at the longer infrared wavelength the gallactic plane became well defined and compact, showing also the bright gallactic center (encircled).

Radioastronomy, originated in the casual observation of radio bursts from the Sun by British and American radar operators in World War II, has become a major source of information on the physical universe in the intervening decades. As remarked by K.I. Kellerman and D.S. Heeschen, the continuous push towards shorter and shorter radio wavelengths combined with vastly increased sensitivity and angular resolution (see Fig. 2.8) has led to important and often unexpected discoveries. Among

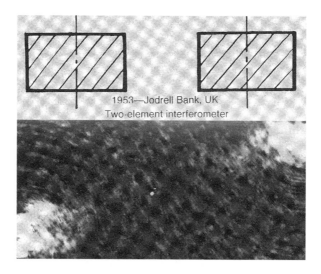

Fig. 2.8. Radio Galaxy Cygnes A: (a) 1953 observation at Jodrell Bank, U.K., showing only that the radio galaxy has two main lobes about 2 arcminutes apart; (b) 1982 observation with the Very Large Array Radio Telescope in New Mexico, U.S.A., showing a central component and hot spots in the well defined lobes.

them, as it is well known are radiogalaxies, quasars, pulsars, interstellar masers, gravitational lenses and the microwave background radiation of 2.7 K discovered by Penzias and Wilson in 1965. At radio wavelengths the perturbations introduced by the atmosphere in telescopic observations can be easily bypassed by means of interferometric techniques, and the angular resolution attainable became much better than that of conventional ground-based optical and infrared telescopes. The most powerful and effective system now in operation is the VLA (Very Large Array) in central New Mexico, which consists of 27 widely separated radio antennas and exceeds in speed, sensitivity, resolution, frequency range, and image size and quality its design specifications. To correct for the incompletely filled antennas and for atmospheric effects very subtle techniques and computer algorithms have been developed. In the near future, VLBA (Very Long Baseline Array), now under construction, will become operative, using an extended baseline which goes from the Caribbean Virgin Islands to distant Hawaii, in the middle of the Pacific Ocean.

The VLBA will use tape recorders and independent hydrogen-maser frequency standards, controlled from the operating center in New Mexico, instead of direct physical connection, which became impractical for such large distances. This system, which will provide by far the highest angular resolution, will be used in studies of active gallactic nuclei, quasars, interstellar measers and radio stars.

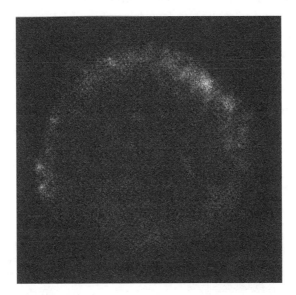

Fig. 2.9. X-ray image of the remnant of Tycho's supernova taken by X-ray satellite ROSAT. The expansion rate of the supernova remnant is several thousand km/sec.

It will also be very useful for the direct determination of distances within our own Galaxy and to nearby galaxies Plans are underway for a MMA (Millimeter Array) facility, with measuring capabilities at millimeter wavelengths exceeding by more of an order of magnitude those of presently used facilities. For the more distant future an international space project between European, Soviet, Japanese and American radioastronomers is being considered which presumably will provide vastly improved sensitivity and image resolution.

The information we receive from the cosmos comes from radiation spanning sixteen orders of magnitude (10^{16}) in wavelength, from radio to gamma rays. But only radio and optical wavelengths can penetrate the atmosphere and reach the Earth, and even these are distorted and perturbed by atmospheric action. Therefore Space Astronomy provides a unique opportunity for observation of the cosmos at large through new windows. Since the 1970's space missions exploring X-ray and gamma-ray emission from the cosmos have provided some evidence for such processes as accreting neutron stars, possible black holes in close binary systems, shock heated remnant of supernova explosions (see Fig. 2.9), hot interstellar and intracluster plasma (see Fig. 2.10) and other instances of highly energetic processes from stars, quasars and unknown sources. In the short period 1988–1991 several important space satellite missions have been lunched, including COBE, HST (Hubble Space Telescope), ROSAT (German-British-American X-ray satellite). In spite unsuspected failures in one of the cases, the mirror flaws discovered in the HST, which will prevent high resolution images from faint objects, all those missions are providing beautiful and precise data, and a wealth of information, which, no doubt, will keep astronomers busy, and the general public wondering for the coming decades.

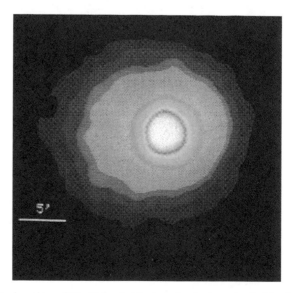

Fig. 2.10. Central region of the Perseus cluster of galaxies revealed by image of X-ray emitting gas taken from space by Einstein Observatory. The scale indicates 5 arc minutes corresponding to 400.000 light years.

Bibliography

1. S.L. Jaki, "The Road of Science and the Ways to God", note 66 to Chap 12, p. 404 (Chicago: University of Chicago Press, 1978).
2. S.L. Jaki, "The Relevance of Physics", p. 244 (Chicago: University of Chicago Press, 1966), and references therein.
3. J.C. Maxwell, "Introductory Lecture in Experimental Physics", Sci. Papers, II, 244 (1871).
4. A. Schuster, "The Progress of Physics during 33 Years, 1875–1808", p. 7 (Cambridge: Cambridge University Press, 1954).
5. "The Nichomachean Ethics of Aristotle", 1094 b, translated by D. Ross, pp. 2–3 (Oxford: Oxford University Press, 1954).
6. S.L. Jaki, "The Relevance of Physics", p. 240 (Chicago: University of Chicago Press, 1966).
7. Ibid., p. 558.
8. Ibid., p. 241.
9. "Sir Isaac Newton's Mathematical Principles of Natural Philosophy and His System of the World", translated by A. Motte; revised and edited with notes by F. Cajori, p. 433 (Berkeley: University of California Press, 1934).
10. P.A. Schilpp (ed.) "Albert Einstein: Philosopher Scientist", p. 100 (Evanston: Library of Living Philosophers, 1949).
11. S.L. Jaki, "The Relevance of Physics", p. 245 (Scottish Academic Press, 1992).
12. "The London Times", November 28, 1919.
13. S.L. Jaki, "The Relevance of Physics", pp. 247–48.
14. Ibid., p. 252.
15. Ibid., pp. 256–57.
16. B. Jaffe, "Michelson and the Speed of Light", p. 168, (New York: Doubleday, Garden City, 1960).
17. E.R. Cohen and B.N. Taylor in "Physics Todays", B.G. p. 9 (August 1990).
18. C.A. Beichman and S. Ridgeway in "Physics Todays", p. 48 (April 1991).
19. J.N. Bahcall, Ibid., p. 24.
20. F.C. Gillet and J.R. Honck, Ibid., p. 32.
21. K.I. Kellerman and D.S. Heeschen, Ibid., p. 40.
22. C.R. Cañizares and B.D. Savage, Ibid., p. 60.

Chapter 3

Masses, Distances and Times in the Universe

3.1. Masses

It is well know that the mass of a star and the mass of a galaxy — the basic units in the universe — are enormous quantities, and also that the mass of the observable universe is still much larger, as a consequence of the fact that the amount of gallaxies observable with the larger telescopes is tantalizingly large. What is perhaps not so well known is that there is a minimum apparent size of a galaxy which is of the order of a few arcseconds.[1] Of course, distant galaxies are receding from us at relativistic speeds, close to the velocity of light, and we are seeing them now as they were several thousands of millions years ago. But the fact remains that the decreasing sizes of more and more distant galaxies stop at the level of a few arcseconds, instead of disappearing gradually, in spite of the fact that sufficiently bright smaller objects could be resolved with existing telescopes. So, the observable universe appears to be finite. On the other hand if things were otherwise, there would be no dark sky at night, as pointed out by Olbers.[2] The reason would be that, if $L(R_m)$ is the light intensity coming from a nearest neighbour galaxy at distance R_m, and the average number of galaxies, is approximately constant throughout cosmic space (assuming large scale homogeneity and isotropy), the total light intensity from a infinite number of galaxies in the universe, L_u, would be given by

$$L_u = \int_{R_m}^{\infty} \bar{n} \frac{L(R_m)}{(R/R_m)} 4\pi R^2 dR = 4\pi R_m^2 L(R_m)\bar{n}[R]_{R_m}^{\infty} \to \infty \qquad (3.1)$$

But, before making reasonable estimates of the mass for a typical star (like our Sun), a typical galaxy (like our Milky Way) and the observable universe, let us remind ourselves what is the mass of the Earth and the mass of the Sun, using simple newtonian arguments. The mass of the Earth can be obtained using the quantity of the gravitational acceleration on its surface $g \approx 980$ cm/sec^2, the quantity of Newton's universal gravitational constant, $G \approx 6.67 \times 10^{-8}$ cm^3 g^{-1} sec^{-2}, and the Earth's radius, $R_T \approx 6378$ km, which is fairly well known since the time of Erathostenes. The values of g and G can be obtained without much difficulty from experiments performed on Earth's surface. We have

$$F = mg = G\frac{mM_T}{R_T^2}, i.e. \ M_T = \left(\frac{g}{G}\right)R_T^2 = 5.97 \times 10^{27} \ g \qquad (3.2)$$

On the other hand the mass of the Sun can be obtained from the fact that, to keep the Earth in its elliptic but very ne a 'v circular orbit around the Sun, it is necessary that the effective centrifugal force be exactly compensated by the gravitational attraction,

$$\frac{M_T(2\pi R_{ST}/T_T)^2}{R_{ST}} = G\frac{M_T M_S}{R_{ST}^2}, i.e. \ M_S = \frac{(2\pi/T_T)^2}{G}R_{ST}^3 \qquad (3.3)$$

$$= 1.99 \times 10^{33} \ g$$

Where R_{TS} is the Earth-Sun distance, and use has been made of the fact that $v_T = 2\pi R_{ST}/T_T$, i.e., the Earth's average velocity, obtained dividing the length of the orbit, $2\pi R_{ST}$, by the period of time necessary to complete one revolution, $T_T = 365 \times 24 \times 60 \times 60$ sec.

In what follows we will show how very simple physical arguments[3] can give us excellent estimates of the mass of stars and galaxies. We will use the dimensionless quantities

$$\alpha \equiv e^2/\hbar c \approx 1/137 \qquad (3.4)$$

and

$$\alpha_g \equiv Gm_p^2 / \hbar c \simeq 6.3 \times 10^{-39} \qquad (3.5)$$

which measure, respectively, the strength of the electrostatic interaction, and the much smaller gravitational interaction, in terms of physical constants (see table in Sec. 2.2). We will also make use of the virial's theorem,[4] which states that for a system in equilibrium in a central field,

$$E_{\text{kin}} = \frac{1}{2} E_{\text{pot}} \qquad (3.6)$$

where E_{kin} = kinetic energy, E_{pot} = potential energy. A star is made up of a plasma of electron and ions (basically protons in the hydrogen burning phase of the life of a star, which is the longest one) and we can equate the kinetic energy of particles in the surface of this plasma which we assume to have spherical shape to a representative electrostatic ionization energy, so that

$$\left(E_{\text{kin}} \right)_S \approx k_B T \approx \frac{e^2}{\alpha r_C} \qquad (3.7)$$

The gravitational potential energy of a proton with mass m_p, on the other hand, is given by

$$\left(E_{\text{pot}} \right)_S \approx G \frac{m_p M_S}{R_S} \qquad (3.8)$$

Then, making use of the virial theorem, Eq. (3.2), we can write, dividing by $\hbar c$ both sides of the equation,

$$\frac{1}{\alpha} \left(\frac{e^2}{\hbar c} \right) \simeq \frac{1}{2} \left\{ \frac{Gm_p^2}{\hbar c c} \right\} \frac{N_S}{R_S / r_C} \qquad (3.9)$$

where N_S is the number of protons within the sphere of radius R_s, and, therefore

$$N_S \simeq \frac{4\pi}{3} R_S^3 \bigg/ \frac{4\pi}{3} r_C^3, \text{ i.e. } (R_S / r_C) \simeq N_S^{1/3} \qquad (3.10)$$

Substituting $(R_S / r_S) = N_S^{1/3}$ in Eq. (3.9), and taking into account the definitions in Eq. (3.4) and (3.5), we get

$$N_S \simeq \left(2\frac{1}{\alpha_g} \right)^{3/2} \simeq 5.65 \times 10^{57} \qquad (3.11)$$

which is the number of protons (baryons) in the plasma sphere which makes up the star. The star's mass, hence, will be given by

$$M_S = N_S m_p \simeq 9.4 \times 10^{33} \text{ g} \qquad (3.12)$$

which is in fairly good agreement with the solar mass evaluated above, Eq. (3.3). Note that the expression for the mass of a typical star as given by Eqs. (3.11) and (3.12) involves only general physical constants, like the combination $\alpha_g = Gm_p^2/\hbar c$ and nip, a truly remarkable result. Figure 3.1 depicts the equilibrium configuration of a star.

The mass of a protogalaxy, i.e. a compact galactic mass presumably in the very first moment after formation, can be estimated in a similar way. In this case gravitational pressure can be assumed to be compensated by radiation pressure from the heated up plasma, and one can conceive a near equilibrium situation when the gravitational free fall time, t_g, for particles in the surface of the quasi-spherical matter distribution, is comparable[5] to the characteristic dissipation time, t_d, of radiant energy coming away from the hot plasma. If $t_g \ll t_d$, the protogalactic plasma would collapse towards the center of mass before reaching quasi-equilibrium conditions. If, on the other hand, $t_g \gg t_d$, radiation pressure would push the protogalactic matter away from the center.

The gravitational free fall time can be given in terms of the protogallactic mass M_G, (which must be the same as the gallactic mass), and the protogallactic radius R_{PG}, (which may be much smaller than the future galactic radius), taking into account the newtonian expression for uniformly accelerated motion,

$$R_{PG} \simeq \tfrac{1}{2}(GM_G/R_{PG}^2)t_G^2 \qquad (3.13)$$

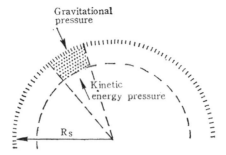

Fig. 3.1. Equilibrium configuration of a star leading to an estimate of its mass.

$$M_S \approx (2/\alpha_g)^{3/2} m_p \approx M_{sun}$$

from which t_g can be eliminated as

$$t_g \simeq (2R_{PG}^3 / GM_G)^{1/2} \qquad (3.14)$$

The characteristic dissipation time can be estimated dividing the total radiation energy, E_r, enclosed by the protogallactic sphere, by the rate at which radiation energy is given up at its surface, i.e.,

$$t_d \simeq \frac{E_r}{c4\pi R_{PG}^2(\sigma T^4)} \simeq \frac{1}{3}\frac{R_{PG}}{c} \qquad (3.15)$$

where as given by Stefan–Boltzmann's law, the energy density per unit volume in terms of the absolute temperature T, is equal to (σT^4), and, therefore

$$E_r = \left(\frac{4\pi}{3}R_{PG}^3\right)(\sigma T^4) \qquad (3.16)$$

Equating t_g and t_d, and taking into account that $M_G = N_G\, m_p$, where N_G in the number of protons in the whole protogalaxy, one gets

$$\frac{2R_{PG}^3}{GN_Gm_p} \simeq \frac{1}{9}\frac{R_{PG}^2}{c^2},\ i.e.\ N_G \simeq \frac{18R_{PG}}{\left(\dfrac{Gm_p^2}{\hbar c}\right)}\left(\frac{m_pc}{\hbar}\right) \qquad (3.17)$$

Now we can make use of the fact that

$$\left(\frac{m_p c}{\hbar}\right) = \frac{m_p}{m_e}\left(\frac{m_e c}{\hbar}\right) = \frac{m_p}{m_e}\frac{1}{\alpha r_c} \tag{3.18}$$

and taking into account that

$$\left(R_{PG}/r_c\right) \approx N_G^{1/3} \tag{3.19}$$

a relationship fully analogous to that in Eq. (3.10), we can finally get

$$N_G \simeq (18 m_p/m_e)^{3/2}\left(\frac{1}{\alpha \alpha_g}\right)^{3/2} \simeq 3.3 \times 10^{67} \tag{3.20}$$

the number of protons (baryons) in the whole protogalaxy. The galaxy mass is, therefore

$$M_G = N_G m_p \simeq 5.5 \times 10^{43}\ \text{g} \tag{3.21}$$

This is of the order of 10^{10} typical star masses, since, as shown in Eq. (3.12), $M_s \approx 9.4 \times 10^{33}$ g. Again, the expression of M_G as given by Eqs. (3.20) and (3.21) involves only numerical factors of order unity and combinations of general physical constants, such as $(m_p/m_c) \approx 1836$, the mass ratio of proton and electron, $\alpha = e^2/\hbar c \approx 1/137$ (dimensionless), $\alpha_g = G m_p^2/\hbar c \approx 6.3 \times 10^{-39}$ (dimensionless), and m_p, the proton mass. It should be mentioned that Eq. (3.21) agrees fairly well with estimates of the total mass of our own galaxy, the Milky Way or Via Lactea, which contains about 10^{11} stars, and therefore has a mass

$$M_{VL} = 10^{11} M_{SUN} \approx 2 \times 10^{44}\ \text{g} \tag{3.22}$$

The mass of the observable universe, M_U cannot be calculated, at least for the time being, in a way similar to those used to estimate M_s and M_G, because, as we know, the whole universe is expanding at fantastic speed, and we are in no way near an equilibrium situation. However one can make a rough estimate of its total mass if one takes seriously current observational estimates of its present average density, ρ_0, its present radius, $R_0 \approx \tau_0/C$, and its spatial curvature $k < 0$, corresponding to an open universe with less mass than that necessary for a hypothetical future phase of contraction after the present expansion.

Using, for instance,

$$\rho_0 \simeq 2 \times 10^{-31} \text{ g/cm}^3 \tag{3.23}$$

$$R_0 \simeq 1.6 \times 10^{28} \text{ cm } (\tau_0 \simeq 1.75 \times 10^{10} \text{ years}) \tag{3.24}$$

one can substitute ρ_0, and R_0 in the expression of M_U to get

$$M_U \simeq \rho_0 V_0 \simeq \rho_0 \frac{4\pi}{3} R_0^3 \simeq 3.8 \times 10^{55} \tag{3.25}$$

which is of the order of 10^{12} galactic masses, equivalent roughly to the actually observable number of galaxies. It may be noted that, strictly speaking, the expression used for the volume

$$V_0 \simeq \frac{4\pi}{3} R_0^3 \tag{3.26}$$

corresponds to the euclidean volume (flat space $|k| = 0$), somewhat different from the more general relationship[7]

$$V_0 \simeq \left[\int_0^1 \frac{4\pi r^2 dr}{(1+|k| r^2)^{1/2}} \right] R_0^3 \tag{3.27}$$

which would be the correct general relativistic expression for a non-euclidean open space. However, the definite integral within square brackets, which is elementary and can be found in tables, becomes $(4\pi/3)$ if $|k|^2 = 0$, and does not differ by more than 15% from this value for $|k| = +1$, or $|k| = -1$, corresponding, respectively, to a closed and an open universe. Errors in the estimates of ρ_0 and R_0 are larger than 15%. The results thus obtained for number of baryons and masses of succesively larger objects in the cosmos are summarized in Table 3.1.

Table 3.1

Object or System	Calculated number of barions	Mass (gr)	Number of units in larger object
Star	$N_s \approx 5.6 \times 10^{57}$	$M_s \approx 9.4 \times 10^{33}$	$(M_G/M_s) \approx 10^{10}$
Galaxy	$N_G \approx 3.3 \times 10^{67}$	$M_G \approx 5.5 \times 10^{43}$	$(M_U/M_G) \approx 10^{12}$
Observable Universe		$M_U \approx 3.8 \times 10^{55}$	$(M_U/M_U) \approx 1$

3.2. Distances

As mentioned in the historical introduction, the oldest estimates of astronomical distances go back to the time of Erathostenes and Aristarchus of Samos. While Erathostenes, using a very simple method, made a very good estimate of the Earth's radius, within of 5% the correct value, Aristarchus went from the Earth to the moon and from there to the sun, in two steps, so to speak. The latter's arguments were correct, but some of the observational data he used lacked the necessary accuracy, and, consequently, his numerical estimates for sizes and distances to moon and sun were seriously flawed. We can reproduce the basic reasoning of Erathostenes and Aristarchus and use modern observational data to obtain successively the Earth's diameter, the distance to the moon and the distance to the sun.

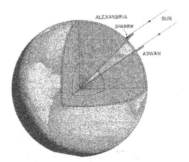

Fig. 3.2. Erathostenes method to measure the Earth's circumference. The distance from S (Syene, Southern Egypt) to A (Alexandria) was $\overline{SA} \approx 5000$ stades ≈ 800 km, and the angle between the vertical and the direction of sunrays at A, on the particular day and time at which no shadow was present at S, could be measured as a = 7.2°. Therefore $2\pi R_T = (\alpha/360) \ \overline{AS} \approx 40,000$ km.

It may be noted that the greek geometer had the good luck of relaying on the distance between two cities, Syene, directly under the ecliptic in Southern Egypt, and Alexandria, in the Mediterranean cost, connected by the Nile river, the only great river running South to North almost along a meridian line in the Northern Hemisphere. Thanks to this happy coincidence he was able to measure with very good accuracy the distance between both cities, 5000 stades. The angle made by sunrays with the vertical at Alexandria (in the same year's day and time in which no shadow could be seen at noon at Syene) was 7.2°, i.e. one fiftieth of a full circle, and consequently, the Earth's circumference was estimated as 250,000 stades or 40,000 km.

Figure 3.3 shows another happy coincidence, perspicaciously noted by Aristarchus, which paved the way for an estimate of the size of the diameter of the moon's orbit. As he noted, it takes, in a lunar eclipse, almost the same time for the moon to begin to reappear after full

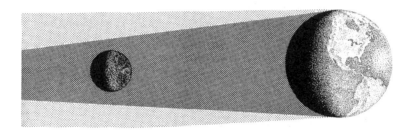

Fig. 3.3. Lunar eclipse. As noted by Aristarchus the path of the moon within the cone of full shadow produced by the Earth is just about two lunar diameters. This fact pointed the way to an estimate of the Earth-moon distance in terms of the lunar diameter.

occultation as it took her to disappear after making contact with the shadow cone produced by the Earth. This implies that the length of the path travelled during the eclipse was about two lunar diameters (more exactly n = 1.8 times the lunar diameter). On the other hand the maximum time for a lunar eclipse is t_{EL} = 1 hour, 42 min. i.e. 102 min, relatively small in comparison with the time needed for the noon to complete a full orbit around the Earth, which is T_L = 27.3 days. Therefore, assuming that the orbital velocity is constant, we have

$$\frac{t_{EL}}{T_L} = \frac{nD_L}{2\pi D_{TL}} =, i.e.$$

$$\frac{d_{TL}}{D_L} = \left(\frac{27.3 \times 24 \times 60}{102}\right)\left(\frac{1.8}{2\pi}\right) \simeq 110.45 \qquad (3.28)$$

where D_L is the lunar diameter and d_{TL} the Earth–moon distance. On the other hand the apparent angular diameter of the moon as seen from the Earth's surface is directly measurable, and is given by

$$\theta_L = \frac{D_L}{d_{TL} - \frac{1}{2}D_T} = 9.205 \times 10^{-3}\,\text{rad}$$

$$i.e.\ \frac{d_{TL}}{D_L} = \frac{1}{\theta_L} + \frac{1}{2}\frac{D_T}{D_L} \tag{3.29}$$

where D_T is the Earth's diameter. Combining Eqs. (3.28) and (3.29) one gets

$$\frac{D_T}{D_L} = 0.276 \tag{3.30}$$

which means that the moon's linear diameter in fairly large, between one forth and one third of the Earth's linear diameter. Also, introducing Eq. (3.30) into Eq. (3.28), one gets

$$\frac{d_{TL}}{D_L} = 30.46 \tag{3.31}$$

which, taking into account that $D_T = (40,000/\pi)$ km leads to $d_{TL} = 384 \times 10^6$ m as the Earth–moon distance.

Finally, to estimate the Earth-sun distance in terms of the Earth-moon distance, Aristarchus considered the triangle made up by Earth, moon and sun when exactly half moon is viewed from the Earth. The angle a subtended from the center of the Earth to the centers of moon and sun is very approximately the same, although slightly smaller, than the corresponding angle subtended from the point nearest the moon on the earths surface (see Fig. 3.4). While it is only eight arc minutes (8') smaller than a right angle, due to the very large distance from sun to Earth, this difference was not beyond the observational capabilities even at Aristarchus time but he did measure it with a large error.

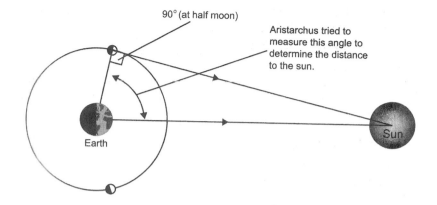

Fig. 3.4. Triangle made up by Earth, Moon and Sun when exactly half moon is viewed from the Earth, which allows an estimate of the Earth–sun distance, d_{TS}, in terms of the Earth–moon distance, d_{TL}, and the angle α.

From Fig. 3.4 it can be seen that

$$\cos \alpha = \frac{d_{TL}}{d_{TS}} \tag{3.32}$$

and, taking into account that the observable value of α is 89 degrees and 52 minutes, very close to a right angle, one gets

$$d_{TS} = \frac{d_{TL}}{\cos \alpha} = 149 \times 10^9 \text{ m} \tag{3.33}$$

a very large distance, indeed.

Also, taking into account that the angular dimension of the sun is close to that of the moon, it may be expected that the linear diameter of the sun be about (d_{TS}/d_{TL}) times the angular diameter of the moon. Using previous results and the fact that

$$\theta_s = \frac{D_S}{d_{TS} - \frac{1}{2} D_T} \simeq \frac{D_S}{d_{TS}} \simeq 12.88 \times 10^{-3} \text{ rad} \tag{3.34}$$

one can readily check that

$$\frac{D_S}{D_T} = 109.1 \tag{3.35}$$

i.e. that the size of the sun is more than one hundred times that of the Earth, or $R_S = R_{SUN} = 6.96 \times 10^5$ km.

These were the first steps of man into its cosmic neighbourhood but, as we will see immediately, these large distances were puny in comparison with the far vaster distances between stars and between galaxies.

The radius of a typical star can be calculated from Eq. (3.10) as

$$R_s = N_S^{1/3} r_c \simeq 8.9 \times 10^5 \text{ Km} \qquad (3.36)$$

where $N_S \approx 5.65 \times 1057$ and $r_c \equiv (1/\alpha)(\hbar/m_e c) = 0.5 \times 10^{-8}$ cm have been used.

This is very close to R_{SUN} as given by Eq. (3.36). The size of a typical star remains fairly constant during most of its long lifetime. On the other hand the same argument applied to a protogalaxy would give, using now $N_G \approx 3.3 \times 10^{67}$, something like

$$R_G = N_G^{1/3} r_c \simeq 1.6 \times 10^9 \text{ Km} \qquad (3.37)$$

which is far smaller than the radius of an actual galaxy of comparable mass. Fortunately we can use observational data on the rotation speeds of outermost stars in an actual spiral galaxy (Andromeda) to get a realistic value for the present radius of a typical spiral galaxy, R_G. Notwithstanding the dark matter (or missing mass) question, it can be argued that the centrifugal acceleration of the outermost star is compensated by the acceleration of gravity towards the galactic center, and, therefore

$$\frac{M_S v_S^2}{R_G} = G \frac{M_S M_G}{R^2_G}, i.e.$$

$$R_G = \frac{G M_G}{v_S^2} = 9.1 \times 10^{21} \text{cm} = 9.1 \times 10^{16} \text{Km} \qquad (3.38)$$

where $G = 6.67 \times 10^{-8}$ cm^{-1} g^{-1} sec^2, $M_G = 5.5 \times 10^{43}$ g and $v_s = 200 \times 10^5$ cm/sec have been used. This is equivalent to

$$\frac{R_G/c}{365 \times 24 \times 60^2} = 9.17 \times 10^3 \text{ light year} \qquad (3.39)$$

which is of the order of the average linear dimension of an actual galaxy. With this information at hand we can estimate the average distance between stars is a galaxy, d_{ss}, which should be roughly equal to the distance from the sun to its nearest neighbour stars. Taking into account that

$$M_G/M_S = \left(\frac{4\pi}{3} R_G^3 \bigg/ \frac{4\pi}{3} d_{SS}^3 \right) \approx 0.58 \times 10^{10} \qquad (3.40)$$

one can write the average distance between nearest neighbouring stars as

$$d_{ss} \approx \frac{R_G}{\left(M_G/M_S \right)^{1/3}} \approx 5.3 \text{ light years} \qquad (3.41)$$

The distance from our Sun to Próxima Centauri,[9] one of its nearest neighbours, with and observable parallax of 0.78 arcseconds, or equivalently 0.78 parsec, is, taking into account that 1 parsec equals 3.26 light years,

$$d \text{ (Sun–Proxima Centauri)} \approx 2.4 \text{ light years} \qquad (3.42)$$

This very large distance explains why, even with the aid of medium size telescopes nearby, stars still appear like geometrical points to the bserver.

Similar considerations lead to an estimate of the distance between neighbouring galaxies. Here we can use

$$M_U/M_G = \left(\frac{4\pi}{3} R_U^3 \bigg/ \frac{4\pi}{3} d_{GG}^3 \right) \approx 0.69 \times 10^{12} \qquad (3.43)$$

and therefore

$$d_{UU} \approx \frac{R_U}{\left(M_U/M_S \right)^{1/3}} \approx 1.8 \times 10^{12} \approx 2.0 \times 10^6 \text{ light years} \qquad (3.44)$$

Galaxies tend to cluster, and recent observations may indicate that they are more irregularly distributed throughout cosmic space than previously thought, but, for comparison, we can give the distance to the nearby galaxy Andromeda, in which the globular cluster M31, containing cepheid variables, serves as distance indicator. The estimated distance from the Milky Way to Andromeda[10] is

$$d_{VL-A} \approx 460.000 \text{ parsecs} = 1.49 \times 10^6 \text{ light years} \qquad (3.45)$$

This fact may be taken as indicative of the reasonableness of the value obtained above for d_{GG}.

Table 3.2 gives a summary of estimated sizes and distances in the universe from stars, through galaxies to the observable universe, as

Table 3.2

Object or system	Radius (cm)	Mass (gr)	Average distance to nearest companion	Density (g/cm³)
Star	8.9×10^{10}	9.4×10^{33}	$d_{ss} \approx 5.0 \times 10^{18}$ cm $\approx 5.3 \times$ l.year	3.2
Galaxy	9.1×10^{21}	5.5×10^{43}	$d_{GG} \approx 1.8 \times 10^{24}$ cm $\approx 2.0 \times 10^6$ l.year	5.8×10^{-24}
Observable Universe	1.6×10^{28}	3.8×10^{55}		2.0×10^{-31}

well as the respective typical densities obtained by means of

$$\rho = M \Big/ \left(\frac{4\pi}{3} R^3 \right) \qquad (3.46)$$

For comparison we can give observational estimates for the mean density of the Sun, the Milky Way[11] and the observable Universe as

$$\rho_S = \rho_{SUN} \approx 1.4 \text{ g/cm}^3 \qquad (3.47)$$

$$\rho_G = \rho_{VL} \approx 3.5 \times 10^{-24} \text{ g/cm}^3 \qquad (3.48)$$

and

$$\rho_U = \rho_0 \approx 2.0 \times 10^{-31} \text{ g/cm}^3 \qquad (3.49)$$

Figure 3.5 gives a plot of log (M) vs. log (R), where M is the mass and R is the radius for different astronomical objects or systems in different states of matter, going from pulsars (very dense neutron stars) to various kinds of stars (white dwarfs, sun-like stars, red giants) to "quasars", which might be akin in mass and density to the protogalaxies discussed above.

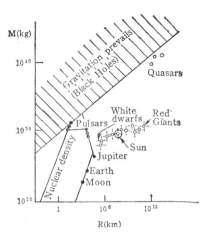

Fig. 3.5. Log-log plot showing M (mass) vs. R (radius) for various objects and systems in the universe.

It is interesting to note that the pulsars, compact neutron stars, have nuclear densities, i.e.

$$\rho_P = \rho_{\text{nuclear matt}} \approx m_p/(4\pi/3)r_p^3 \approx 10^{14} \text{ g/cm}^3 \qquad (3.50)$$

The Moon, the Earth, and Jupiter, having comparable densities, fall more or less in line with each other. The Sun, on the other hand, in comparison with the smaller stars (white dwarfs) and the larger ones (red giants) differs by a moderate factor in mass but by a much lager factor in size. Quasars have of the order of 10^{10} solar masses but are much more compact than normal galaxies (not shown in the graph).

3.3. Times

About the time evolution of stars some progress has been made[13] in recent decades, but no small amount is still to be learned. Less is known about the time evolution of galaxies, and still less about galaxy clusters.

Some of the stars in globular clusters may be almost as old as the universe itself. Other stars, like our sun, may be second or third generation stars, and, therefore, much younger. When a second generation star begins to condense out of cosmic dust, thrown away into interstellar space by previously exploding first generation stars, it at first shines with intense red light and then contracts, decreases in luminosity, and changes the color (wavelength) of the emitted light towards the green. This means that it is approaching the quasi-equilibrium state in which the inner gas pressure is balanced by the gravitational pressure of the spherical mass distribution. At this point fussion processes, converting light hydrogen to heavier ^{4}He are taking place in the central nucleus of the star. We know that the sun is not a first generation star because it contains considerable amounts of heavy elements in addition to hydrogen and helium. The process of nuclear fusion is a very efficient one in producing energy, and the conditions under which this process takes place in the hugh nuclear reactor which is the stellar interior are such that hydrogen burning proceeds at a slow pace, releasing quietly a very large amount of energy per unit time.

We can make a rough estimate of the life-time span of an average star, like our sun, taking into consideration the amount of available nuclear fuel, and the rate at which this fuel is burnt away and projected to space in all directions in the form of light, or electromagnetic radiation. First, let us assume that only a certain fraction of the total hydrogen supply is used during the life-time of an average star. We may assume this fraction to be fifty percent, or

$$f = 0.50 \tag{3.51}$$

Second we must take into account that only a fraction of the mass of the fussed constituents is converted to energy, in a fussion process. This fraction is the binding energy[14] per nucleon (see Fig. 3.6).

The information in Fig. 3.6 tells us that for every fussion process ending up in one ^{4}He nucleus we get out an energy B $\approx$ 7 MeV per nucleon, i.e. the useful energy conversion fraction is

$$\frac{B\left(^4 He\right)}{m_p c^2} = \frac{7 \times 10^6}{1.67 \times 10^{-24}\left(3 \times 10^{10}\right)^2 / 1.6 \times 10^{-12}} = 7.2 \times 10^{-3} \tag{3.52}$$

where the conversion factor 1 erg = $(1/1.6 \times 10^{-12})$ eV has been used.

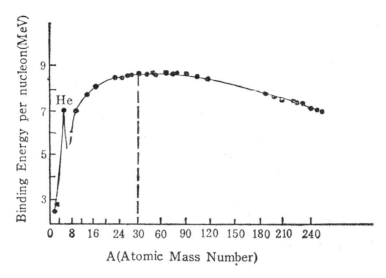

Fig. 3.6. Average energy per nucleon (MeVsinucleon) for stable nuclei as a function of nuclear mass (atomic units) or total number of nucleons, in nuclei of increasing mass.

Taking into account that for a star of radius R_s and surface temperature T_s the energy radiated per unit time into outer space is, according to Stefan-Boltzmann's radiation law,

$$\text{Rate of energy radiation} = c[4\pi R_S^2 \sigma T_S^4] \qquad (3.53)$$

we can write

$$\frac{d}{dt}\left[f\left(\frac{B(^4He)}{m_p c^2} \right) M_C^2 \right] = -c[4\pi R_s^2 \sigma T_s^4] \qquad (3.54)$$

and, making use of

$$M = \frac{4\pi}{3} R_S^3 \rho_S \qquad (3.55)$$

where ρ is the density, we can integrate from $t = 0$ to $t = T_s$. Here τ_s is the time at which the useful fuel supply is exhausted, and we may assume that the star has collapsed to a size half the original size, just before exploding. The result of the integration is

$$\tau_S = \frac{f\left[\left(B\left(^4He\right)/m_P c^2\right)\right]}{\sigma T_S^4}\left[R_S(0) - R_S(\tau_S)\right] \tag{3.56}$$

We can substitute here f, from Eq. (3.51), $[B(^4He)/m_p c^2]$, from Eq. (3.52), the constants $c = 3 \times 10^{10}$ cm and $\sigma = 7.63 \times 10^{-15}$ erg cm^{-3} K^{-4}, and appropriate values, like those pertaining to a typical star (our sun) given by

$$\rho_S = 1.41 \text{ g/cm}^{-3} \tag{3.57}$$

$$R_S = 0.696 \times 10^{11} \text{ cm}; \ R_S(0) - R_S(\tau_S) \approx (\tfrac{1}{2})R_S(0) \tag{3.58}$$

$$T_S \approx 6000 \text{ K} \tag{3.59}$$

to get

$$\tau_s = \tau_{SUN} \approx 5.2 \times 10^{17} \text{ sec} = 1.68 \times 10^{10} \text{ years} \tag{3.60}$$

This is a reasonable value for a star of the mass and size of the sun. More massive stars have much higher surface temperatures and much shorter lifetimes. We may tray putting it in a semiquantitative way in a moment.

The temperature in the interior of a star, T_s^*, much higher than its surface temperature, T_s, is given by

$$k_B T_S^* \approx \frac{GM_s m_P}{R_S} \tag{3.61}$$

through the virial's theorem, where k_B, is Boltzmann's constant, G is Newton gravitational constant and m_p the proton mass. For our sun, which has a mass $M_s \approx 2 \times 10^{33}$ g and a radius $R_s \approx 6.96 \times 10^{10}$ cm, we get

$$T_{SUN}^* = \frac{GM_{SUN}m_P/k_B}{R_{SUN}} \approx 2.32 \times 10^7 \text{ K} \tag{3.62}$$

or more than twenty million degrees Kelvin, which is much higher than its surface temperature,

$$T_{SUN} \approx 6000 \text{ K} \tag{3.63}$$

From Eqs. (3.61) and (3.62) we may conclude that the central temperature of a star of mass M_s and radius R_s is related to the central temperature of the sun by

$$\frac{T_S^*}{T_S^*} = \frac{M_S/M_{SUN}}{R_S/R_{SUN}} \qquad (3.64)$$

i.e., knowing the mass and the radius of the star, we can estimate the temperature in its interior.

In what follows we will make a semiquantitative estimate of the lifetime of a star having between 0,3 and 30 solar masses.

The absolute luminosity or rate of energy radiation per unit time is given, according to Eq. (3.53), by

$$L = c\,4\pi R_S^2 \sigma T_S^4 \qquad (3.65)$$

For the Sun this is

$$L_{SUN} = c\,4\Pi R_{SUN}^2 \sigma T_{SUN}^4 \qquad (3.66)$$

and therefore

$$\frac{R_S}{R_{SUN}} = \left(\frac{L_S}{L_{SUN}}\right)^{1/2} \left(\frac{T_{SUN}}{T_s}\right)^2 \qquad (3.67)$$

where both the luminosity and the surface temperature are in principle observable quantities (Hertzsprung–Russel diagram). Knowing the mass and radius of a star relative to those of the sun it is straight forward to get the relative density,

$$\frac{\rho_S}{\rho_{SUN}} = \frac{M_S/M_{SUN}}{(R_S/R_{SUN})^3} \qquad (3.68)$$

Using the known mass, surface temperature and absolute luminosity of a star, information about its radius, by means of Eq. (3.67), its density, Eq. (3.68), and its inner temperature, Eq. (3.64) can be easily obtained, as given in Table 3.3 all quantities in the table are referred

Table 3.3. Properties of stars.

Star	M_S/M_{SUN}	T_S/T_{SUN}	L_S/L_{SUN}	R_S/R_{SUN}	ρ_S/ρ_{SUN}	T_S^*/T_ρ^*
Small star	0.3	0.6	0.02	0.4	15.	0.77
Average star	1.0	1.0	1.0	1.0	1.0	1.0
Large star	30.0	5.3	3×10^5	19	1.0×10^{-4}	1.56

to the corresponding solar quantity, i.e.

$$M_{SUN} \approx 2 \times 10^{33} \text{ g} \tag{3.69}$$

$$T_{SUN} \approx 6000 \text{ K} \tag{3.70}$$

$$L_{SUN} \approx 18 \times 10^{34} \text{ erg/sec} \tag{3.71}$$

$$R_{SUN} = 6.96 \times 10^{10} \text{ cm} \tag{3.72}$$

$$\rho_{SUN} = 1.41 \text{ g/cm}^3 \tag{3.73}$$

$$T_{SUN}^* = 2.32 \times 10^7 \text{ K} \tag{3.74}$$

Finally we can make use of Eq (3.56) to estimate the lifetimes (in the hydrogen burning or main sequence phase, which is by far the longest phase) of stars ranging in mass from 0,3 M_{SUN} to 30 M_{SUN}. According to this equation we have, in terms of solar quantities

$$\frac{\tau_s}{\tau_{SUN}} = \frac{\left(\rho_s/\rho_{SUN}\right)\left(R_s/R_{SUN}\right)}{\left(T_s/T_{SUN}\right)^4} \tag{3.75}$$

where we know, by Eq. (3.60), that the lifetime for the sun in the main sequence is about 10^{10} years. Results are given in Table 3.4. We can see that small star, with thirty per cent the mass of the sun, has a

Table 3.4. Lifetime of stars.

Star	Mass (M_S/M_{SUN})	Luminosity (L/L_{SUN})	Approx. lifetime (years)
Small star	0.3	0.02	10^{11}
Average star	1	1.	10^{10}
Large star	30	3×10^5	10^6

much lower luminosity, and therefore it lasts longer than the sun. Our sun, an average mass star, has already lived for about 0.5×10^{10} years and, after completing its life span in another 0.5×10^{10} or so, it will become[15] a red giant with 1000 times its present radius, moving up away from the main sequence in the Hertzsprung-Russell diagram. Eventually, after exhausting its residual nuclear fuel, will become a white dwarf, moving to the lower left side of the Hertzsprung-Russell diagram. A really massive, $M = 30\ M_{SUN}$, highly luminous star, spends quickly the hydrogen supply and lives in the main sequence for a period of 10^6 years, or about 10000 times shorter than that of the sun.

In this chapter we have seen that it is possible to go a long way in understanding the scales of masses, lengths and times in the universe. At first sight it may appear an inmodesty on the part of men to try to understand such a thing as the whole universe. Of course we will never understand in its full depth the purely physical universe, in the same way that we do not expect to fully understand a flower, a rabbit, or, much less, a human being. But certainly we are able to understand a fair amount, as witnessed by our ability to quantify such large scales of masses, distances and times as they are needed to describe stars and galaxies. Anatole France once said[16] "the universe which science reveals to us is a dispiriting monotony. All the suns are drops of fire and all the planets drops of mud". As a man made not only of flesh but also of spirit he certainly was superior to the physical universe, but the air of superiority which transpire these words, shows, more than anything else, his very poor appreciation of the truly human adventure of science in general, and the science the universe in particular.

Bibliography

1. C.A. Beichman and S. Ridgway, "Physics Today", p. 49 (April 1991).
2. S.L. Jaki. "The Paradox of Olbers' Paradox" (New York: Herder and Herder, 1969).
3. J. Silk, "Physics Today", p. 28 (April 1987).
4. L.D. Landau and E.M. Lifshitz, "Mechanics", 3rd ed, p. 23 (Oxford: Pergamon Press, 1976).
5. J. Silk "Physics Today", p. 28 (April 1987).
6. Corresponding to $\Omega_0 = p_0/p_c = 0.025$. For further discussion see next chapter.

7. E.P. Tryon, "Cosmic Inflation", p. 717, in "Encyclopedia of Physical Science and Technology". Vol. 3 (New York: Academic Press, 1987).
8. J. Silk, "The Big Bang", p. 190 (San Francisco: W.H. Freeman and Co. 1980).
9. S.L. Jaki "The Relevance of Physics", p. 248 (Chicago: University of Chicago Press, 1966).
10. S. Weinberg, "Gravitation and Cosmology", p. 436 (New York: Wiley, 1972).
11. Fang Li Zhi and Li Shu Xian, "Creation of the Universe", p. 14 (Singapore: World Scientific, 1989).
12. C. Gerthsen, H.O. Kneser and H. Vogel, "Physik", Fig. 17.9 (Berlin: Springer-Verlag, 1977).
13. J. Silk, "The Big Bang" (San Francisco: W.H. Freeman and Co., 1980).
14. R.D. Evans, "The Atomic Nucleus", p. 299 (New York: McGraw Hill, 1955).
15. R. Jastrow, "Red Giants and White Dwarfs" (New York: Warner Books, 1980).
16. A. France, "La vie littéraire", Vol. 3, p. 212 (Paris: C. Lévy, 1888–92) [Cited by S.L. Jaki in "God and the Cosmologists", p. 218 (Washington: Gateway, 1989)].

Chapter 4

Relativistic Cosmology

4.1. Relativity, Special and General

Physics is simple, but subtle, according to Einstein. The General Theory of Relativity is very general, and consequently very subtle. Its complex mathematical language — tensorial analysis — requires special skill and competence to apply it to definite physical problems. Fortunately, however, the apparent simple characteristics of the physical universe sphericity, homogeneity, isotropy — reduce considerably the complexity of the general relativistic equations for the universe at large, and make it possible to get simple solutions in which the analogy with the corresponding non-relativistic newtonian equations can be ascertained.

Fig. 4.1. A. Einstein.

In the special theory of relativity Einstein had established that given an inertial frame of reference, any other frame of reference moving uniformly and without rotation with respect to the first is also an inertial frame of reference, and therefore the laws of nature are concordantly described in the two inertial reference systems (principle of special relativity). In this theory space and time are no longer absolute but depend on the direction and magnitude of the translation speeds of the inertial frame of the observer with respect to the inertial frame where the observed event takes place. This is true, of course, for speed, comparable to the speed of propagation of light. Otherwise, i.e. for speeds, v, such that $(v/c)^2 \ll 1$, relativistic space and time become indistinguishable from absolute space and time. On the other hand the velocity of light in empty space becomes and absolute universal constant, independent of the relative speed of the reference system of the observer, in agreement with the Michelson-Morley experiment. So the theory of relativity is more "absolutist" than generally believed.

But having established, in the special theory of relativity, that physical laws were the same (invariant) regardless of the speed of the inertial (i.e. non-accelerated) reference system from which they were observed, Einstein went a step further, and in his General Theory of Relativity he established that physical laws were also the same regardless of the acceleration of the reference frame. He got a clue to this step from the equivalence of inertial and gravitational mass. An observer momentaneously moving at constant speed would be unable to attribute a change in speed either to a sudden acceleration of his motion or to the sudden appearance of an external gravitational field. Pointing our that newtonian mechanics assumed a "spatium absolutum" and a "tempus absolutum" while special relativity assumed a "continuum spatium-tempus absolutum", he found this still unsatisfactory. This is so because, on one hand, something (the spatio temporal continuum) which acts by itself but cannot be acted upon, is conceptually unreasonable in itself, and, on the other hand, because, for Einstein, the spirit of the principle of relativity required an extension to non-inertial frames of reference, or otherwise, such experimental fact as the observational equivalence between inertial and gravitational mass would not be justifiable. The

General Theory of Relativity implies, therefore, that the geometry of space-time is modified by the presence of massive material bodies. The gravitational field thus influences the metrics of space-time and even determines it. Fortunately, first Gauss, introducing arbitrary curvilinear coordinates adapted to the geometry of any two dimensional surface and, later, Riemann, generalizing this idea to more than two dimensions, had prepared the ground for the development of tensorial calculus, which is the mathematical tool required by the General Theory of Relativity.

As pointed out previously, experimental tests of general relativity include[2] observations on the deflection of light by the sun (and other astronomical objects), the precession of perihelia of planets, and radar echo delays. More refined experiments along these lines are under way, and, for the time being, no conclusive experimental evidence has been found against its validity. Its beauty and generality speak for themselves, and it may be said that a general consensus exists among physicists in its favour. On the other hand, attempts to make it compatible with quantum theory, which is known to be eminently successful in the atomic and subatomic realm, have not been too rewarding up to now.

4.2. The Cosmological Dynamic Equations

Einstein's field equations[3] for a spherical, homogeneous and isotropic system (Robertson-Walker metrics) reduce to

$$\dot{R}^2 = \frac{8\pi}{3}G\rho R^2 - kc^2 \qquad (4.1)$$

where R = R (t) is the scale factor or radius, $\dot{R} = \dot{R}(t)$ its time derivative, G = 6.67 × 10^{-8} dyn. cm^2/g^2 is Newton's gravitational constant, $\rho = \rho(t) = \rho_m(t) + \rho_r(t)/c^2$ the mass density, which involves both matter and radiation mass, k the spatial curvature, which can in principle be positive (closed universe) or negative (open universe), and c is the velocity of light.

It can be noted that for k = 0 (euclidean space) Eq. (4.1) can be written as

$$\frac{1}{2}m\dot{R} = Gm\frac{\left(\frac{4\pi}{3}\rho R^3\right)}{R} = G\frac{mM}{R} \qquad (4.2)$$

which is the classical newtonian equation for the motion of a mass m
moving radially with kinetic energy $(1/2)m\dot{R}^2$ under the influence of a
central gravitational potential enclosing total mass M and therefore with
potential energy *GmM/R*, at escape velocity. As the time goes to infinity
($t \to \infty$) the kinetic and potential energies go simultaneously to zero, because
R $\to$ 0, R $\to$ ∞. Figure 4.2 depicts the expansion (we know that the
universe is expanding because galaxies are receding from each other
according to the Hubble's law, $\dot{R} = HR$) of a spherical homogeneous and
isotropic mass distribution in accordance with Eq. (4.1).

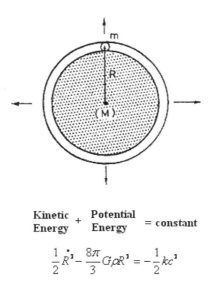

$$\text{Kinetic Energy} + \text{Potential Energy} = \text{constant}$$

$$\frac{1}{2}\dot{R}^2 - \frac{8\pi}{3}G\rho R^2 = -\frac{1}{2}kc^2$$

Fig. 4.2. General relativity dynamical equation for an expanding spherical, homogeneous,
isotropic mass distribution. The spatial curvature k can be k > 0 (closed universe in which
the expansion phase is necessarily followed by a contraction phase because the expansion
takes place at less than scape velocity), k = 0 (flat euclidean universe expanding for ever
at exactly scape velocity) and k < 0 (open universe, expanding for ever at a velocity
higher than scape velocity).

In addition to Eq. (4.1) we can write down the energy conservation equation

$$\frac{d}{dR}(\rho R^3) = -3pR^3 \qquad (4.3)$$

where p is the pressure due to matter and radiation within the spherical mass distribution, and the equation of state

$$p = p(\rho) \qquad (4.4)$$

which is not known explicitly in general, but it is given by $p_r = \rho_r/3$ for radiation pressure only, in particular.

Equations (4.1) (Einstein equation), (4.3) (energy conservation equation) and (4.4) (equation of state) are the fundamental equations of dynamical cosmology.

The critical mass density, p_c, is defined as the mass density necessary to make the metrics to the universe flat (k = 0) and therefore, from Eq. (4.1),

$$\dot{R}^2 = \frac{8\pi}{3}G\rho R^2, \text{i.e. } \rho_c \equiv 3\left(\dot{R}^2/R^2\right)/8\pi G \qquad (4.5)$$

Combining Eqs. (4.1) and (4.5) we get

$$-k = \left(\frac{R^2}{c^2}\right)\left(1 - \frac{\rho}{\rho_c}\right) \qquad (4.6)$$

This means that the sign of the space curvature is entirely determined by the ratio of the actual mass density of the critical mass density in the universe, i.e.

$$\left(\rho/\rho_c\right) > 1 \text{ implies } k > 0 \text{ (closed universe)} \qquad (4.7)$$

$$\left(\rho/\rho_c\right) \le 1 \text{ implies } k \le 0, \ -k = |k| \text{ (open universe)} \qquad (4.8)$$

Present estimates of the mass density, based upon galaxy counts in large regions surrounding our own galaxy and estimates of the typical mass of galaxies deduced from their rotation velocity indicate that $(\rho/\rho_c)<1$, which supports an open universe. Many theoretical cosmologists, however, remain unconvinced and hope that some of the various candidates proposed (from mini black holes to neutrinos) can provide the extra mass density needed to close the universe. This is the so called "missing mass problem".

Assuming that $-k = |k|$ (open universe) we can rewrite Eq. (4.1) as

$$\dot{R} = R^{-\frac{1}{2}}\left\{(8\pi/3)G\rho R^3 + |k|c^2 R\right\}^{\frac{1}{2}} \qquad (4.9)$$

If it is further assumed that after a conveniently unspecified early time the pressure p becomes negligible, and that according to Eq. (4.3),

$$\rho R^3 = \text{constant}, \qquad (4.10)$$

the first term within curly brackets in Eq. (4.9) becomes constant also. Consequently we can define p_+ and R^3_+, whose product is

$$\rho_+ R^3_+ = \rho R^3 = \text{constant}, \qquad (4.11)$$

in such a way that the two terms within curly brackets in Eq. (4.9) become equal,

$$(8\pi/3)G\rho_+ R^3_+ = |k|c^2 R_+, \text{ i.e, } R_+ = \left(3|k|c^2/8\pi G\rho_+\right)^{\frac{1}{2}} \qquad (4.12)$$

and then proceed with the integration of Eq. (4.9), which reduces to

$$\int dt = \int \frac{R^{\frac{1}{2}}}{\left\{(8\pi G/3)\rho_+ R^3_+ + |k|c^2 R\right\}^{\frac{1}{2}}} \qquad (4.13)$$

The integral in the right hand side can be performed making the change of variable

$$x^2 = |k|c^2 R \qquad (4.14)$$

defining,

$$a^2 = (8\pi\, G/3)\, \rho_+ R_+^3 = |k|\, c^2 R_+, \tag{4.15}$$

and taking into account the result for the indefinite integral

$$\int \frac{x^2}{\left\{a^2 + x^2\right\}^{1/2}}\, dx = \frac{x}{2}\left\{a^2 + x^2\right\}^{1/2} - \frac{a^2}{2}\ln\left[x - \left\{a^2 + x^2\right\}^{1/2}\right] \tag{4.16}$$

which can be found in tables and can be checked by direct differentiation.

Using Eqs. (4.14)–(4.16) one gets

$$t = \frac{R_+}{|k|^{1/2}\, c}\left\{\left(\frac{R}{R_+}\right)^{1/2}\left(1 + \frac{R}{R_+}\right)^{1/2} - \ln\left[\left(\frac{R}{R_+}\right)^{1/2}\left(1 + \frac{R}{R_+}\right)^{1/2}\right]\right\}$$

$$= \frac{R_+}{|k|^{1/2}\, c}\left\{\sinh y \cosh y - y\right\} \tag{4.17}$$

where the definition

$$\left(R/R_+\right)^{1/2} \equiv \sinh y \tag{4.18}$$

which implies

$$R = R_+ \sinh^2 y \tag{4.19}$$

has been used. Eqs. (4.17) and (4.19) give the time t and the scale factor R in terms of y in parametric form, and are equivalent to the well known Friedmann[5] solutions for an open universe. For a closed universe (k > 0) the solutions are very similar and are given in terms of the trigonometric functions siny, cosy instead of the hyperbolic functions sinhy, coshy.

The basic parameters to describe the time evolution of the universe are the Hubble parameter (H), the density parameter (Ω) and the deceleration parameter (q), defined by

$$H \equiv \dot{R}/R \tag{4.20}$$

Fig. 4.3. E.P. Hubble.

Fig. 4.4. M.L. Humason.

$$\Omega \equiv \rho / \rho_C = 1 - \frac{|k|c^2}{R^2} \quad \text{(see Eq. (4.6))} \qquad (4.21)$$

$$q \equiv -\ddot{R}R / \dot{R}^2 \qquad (4.22)$$

All these parameters are given in terms of

$$\dot{R} = \left(dR/dy \right) / \left(dt/dy \right), \qquad (4.23)$$

and,

$$R = \left(d\dot{R}/dy\right)\Big/\left(dt/dy\right) \tag{4.24}$$

which can be easily calculated from Eqs. (4.17) and (4.19), resulting in

$$\dot{R} = |k|^{1/2} c \frac{\cosh y}{\sinh y} \tag{4.25}$$

and

$$\ddot{R} = -\frac{1}{2} \frac{|k|^{1/2} c}{R_+} \frac{1}{\sinh^4 y} \tag{4.26}$$

Substituting now these values for $\dot{R}$ and $\ddot{R}$ in Eqs. (4.10)–(4.24) one finally gets

$$H = \left(|k|^h c/R_+\right)\left(\cosh y/\sinh^3 y\right) \tag{4.27}$$

$$\Omega = 1/\cosh^2 y \tag{4.28}$$

$$q = 1/2 \; \cosh^2 y \tag{4.29}$$

We can see that for $R \ll R_+$, i.e. $y \ll 1$, one has

$$H(y \to 0) \approx \left(|k|^{1/2} c/R_+\right) y^{-3} \to \infty \tag{4.30}$$

$$\Omega(y \to 0) \approx 1 \tag{4.31}$$

$$q(y \to 0) \approx 1/2 \tag{4.32}$$

$$H_+ = (0.9507)\left(|k|^{1/2} c/R_+\right) \tag{4.33}$$

$$\Omega_+ = 0.4199 \tag{4.34}$$

$$q_+ = 0.2099 \tag{4.35}$$

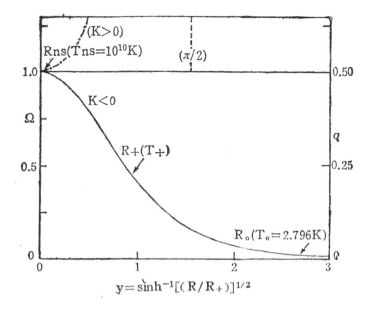

Fig. 4.5. Dependence of $\Omega \equiv \rho/\rho_c$ and $q \equiv -\ddot{R}R/\dot{R}^2$ upon the parameter $y = \sinh^{-1}(R/R_+)^{1/2}$ which is related to the degree of expansion of the universe.

And, finally, for $R \gg R_+$, i.e. $y \gg 1$,

$$H(y \to \infty) \approx \left(|k|^{\frac{1}{2}} c/R_t \right) 4e^{-2y} \to 0 \qquad (4.36)$$

$$\Omega(y \to \infty) \approx 4e^{-2y} \to 0 \qquad (4.37)$$

$$q(y \to 0) \approx 2e^{-2y} \to 0 \qquad (4.38)$$

The dependence of Ω and q on $y = \sinh^{-1}[(R/R_+)]^{1/2}$ is depicted in Fig. 4.5 where it can be seen that both decrease monotonously towards zero for k < 0 as the expansion proceeds. When the expansion is in the very early stages (R ≪ R₊) the value of $\Omega = \rho/\rho_c$ becomes almost the same for k < 0, k = 0, and k > 0, making it difficult to distinguish weather the universe is open, flat or closed.

It is of interest to set upper and lower limits to the present density
$(\Omega) = (\rho_0/\rho_{c0})$ and present age $(\tau_0 = t(y_0)/H_0^{-1})$ of the universe based
upon Eqs. (4.28) and (4.17) and upon the available information on
the minimum mass density estimated from the mass in galaxies,[6]
the minimum age of the universe estimated from the age of the oldest
stars in the Milky Way,[7] and present reasonable values of Hubble's
constant,[8] whose observational value is known within a factor of two
only, i.e. it lays within $0.7\ H_0$ and $1.4\ H_0$, being H_0, its most probable
value.

Assuming $\Omega_0 = 0.025$, $H_0 = 2.1 \times 10^{-18}$ s^{-1}, one can make use of
Eq. (4.28) to get the corresponding value of the parameter y_0 through

$$y = \cosh^{-1}(1/\Omega_0)^{\frac{1}{2}} = 2.53 \qquad (4.39)$$

and then get the corresponding age of the universe by Eq. (4.17)

$$t = \frac{R_+}{|k|^{\frac{1}{2}}c}\left(\sinh y_0 \cosh y_0 - y_0\right)$$

$$= \frac{\cosh y_0/\sinh^3 y_0}{H_0}\left(\sinh y_0 \cosh y_0 - y_0\right) \qquad (4.40)$$

$$= 4.57\times10^{17}\,s = 1.45\times10^{10}\,\text{years}$$

which is very close to the estimated age of the oldest stars in the Milky
Way ($t \approx 1.5 \times 10^{10}$ years).

Figure 4.6 gives $\tau_0 \approx t_0/H_0^{-1}$ as a function of Ω_0 in the interval $10^{-2} <$
$\Omega_0 < 1$ for three different values of the Hubble constant ($0.7\ H_Q$, H_0 and
$1.4\ H_0$) spanning the uncertainly range for this fundamental cosmological
parameter. It can be seen that, while $\Omega_0 \approx 0.025$ is favoured, values as
large as $\Omega_0 \approx 0.7$ are not completely ruled out by other observational
constraints.

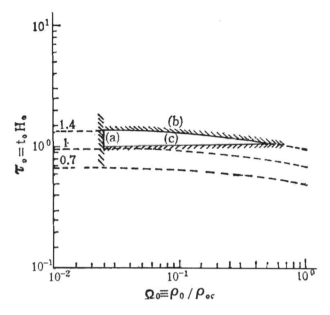

Constraints
(a) Minimum mass in galaxies
(b) Maximum Hubble's constant
(c) Age of oldest stars in Milky way

Fig. 4.6. Limits on the mass density and the age of the universe compatible with cosmological equations (Eqs. (4.26) and (4.17)) for an open (k < 0) universe ($H_0 = 2.1 \times 10^{-18}$ s^{-1}).

4.3. The Matter Dominated and the Radiation Dominated Eras

We know that at present times in the expansion of the universe the mass density associated with matter (ρ_m) is much larger than the mass density associated with background radiation (ρ_r/c^2). The former, which is concentrated in stars and cosmic dust within galaxies, can be estimated as

$$\rho_{m0} \approx 1.97 \times 10^{-31} \text{ g/cm}^3 \qquad (4.41)$$

corresponding to $\Omega_0 = \rho_{m0}/\rho_{c0} = 0.025$, where $\rho_{c0} = 3H^2/8\pi G$, with H_0, taken as $H_0 = 2.1 \times 10^{-18}$ s^{-1}. The mass density associated with background radiation, on the other hand, can be obtained directly introducing the present background radiation temperature, $T_0 = 2.796$ K, which is known with high precision,[9] into the Stefan-Boltzmann relationship,

$$\rho_{r0}/c^2 = \left(\sigma T_0^4\right)\big/c^2 = \left(7.63\times10^{-15}\right)(2.796)^4\big/(3\times10^{10})^2$$
$$= 5.18\times10^{-34} \text{ g/cm}^3 \tag{4.42}$$

Further, we know that at present matter and radiation are "decoupled", because the universe is basically transparent after atoms have been formed and most matter is concentrated in stars and cosmic atomic dust inside galaxies separated by large distances. Accordingly we may assume that the expansion of matter and the expansion of radiation proceed separately, with negligible interaction with one another. Also, due to the fact that at present times the universe as a whole is cool enough, interconversion processes between matter and radiation at a large scale can be ruled out. Then we can use the energy conservation equation, Eq. (4.3), separately for matter and for radiation to determine the evolution of $\rho_m(R)$ and $\rho_r(R)/c^2$ throughout the expansion. For matter we have, at present, $\rho = 0$, and, therefore

$$\frac{d}{dR}\left(\rho_m R^3\right) = 0, \text{ i.e. } \rho_m R^3 = \rho_{m0} R_0^{\,3} \tag{4.43}$$

where ρ_{m0} and R_0 stand for present values, and ρ_m and R for matter density and scale factor at times after "decoupling" and into the foreseeable future.

For radiation, on the other hand, we have $\rho_r = \rho/3$, and, consequently

$$\frac{d}{dR}\left(\rho_r R^3\right) = -3, \; (\rho_r/3)R^2 = -\frac{1}{R}(\rho_r R^3) \tag{4.44}$$

Eq. (4.44) can be easily integrated to get

$$\ln\left(\rho_r R^3\right) = -\ln R, \text{ i.e. } \rho_r R^4 = \rho_{r0} R_0^{\,4} = \text{constant} \tag{4.45}$$

Making use now of the Stefan-Boltzmann relationship, which gives the radiation density in terms of the absolute temperature, we may write

$$\rho_r = \sigma T^4, \quad \rho_{r0} = \sigma T_0^{\,4}, \quad \text{i.e.} \quad \rho_r T^{-4} = \rho_{r0} T_0^{\,4} = \sigma \tag{4.46}$$

which, combined with Eq. (4.46) leads to

$$RT = R_0 T_0 = \text{constant} \tag{4.47}$$

This result, which makes a precise quantitative connection between scale factor, R, and absolute temperature, T, is strictly valid as long as radiation and matter are "decoupled". As will be shown below, together with Eqs. (4.43) and (4.45) it leads directly to the prediction of a constant (and very small) ratio of massive particles (baryons) to photons in the universe, which is usually supposed to extend back to very early times (high temperatures), much earlier than "decoupling". If at temperatures higher than the "decoupling" temperature matter and radiation interact strongly it would be necessary to introduce some friction between matter and radiation which could give rise to a changing ratio of baryons to photons at earlier epochs.

Since, according to Eq. (4.43), ρ_m goes as R^{-3} through the expansion, and according to Eq. (4.45), ρ_r goes as R^{-4}. The ratio $\rho_m/(\rho_r/c^2)$ is increasing with increasing expansion time, and was therefore smaller and smaller at earlier times, until becoming unity at a certain time which is thus defined as the "decoupling" time. Before estimating the "decoupling" temperature, we may note that the auxiliary parameter

$$y \equiv \sinh^{-1}\left(\frac{R}{R_+}\right)^{1/2} \tag{4.48}$$

in terms of which the time, Eq. (4.17), and the scale factor, Eq. (4.19) are given, can be expressed by virtue of Eq. (4.49), in the form

$$y \equiv \sinh^{-1}\left(\frac{T_+}{T}\right)^{1/2} \tag{4.49}$$

which is much more convenient to deal with, because knowing T_0 and y_0 (obtainable from the value of Ω_0) we can determine directly T_+, and therefore calculate t(y) in a straightforward manner for any T of interest, for instance T_d (decoupling), T_{af} (atom formation) or T_{ns} (nucleosynthesis). From $y_0 = 2.53$, corresponding to $\Omega_0 = 0.025$ (see Eq. 4.39) we get

$$T_+ = T_0 / \sinh^2(y_0) = 108.7 \text{ K} \qquad (4.50)$$

which comes out to be considerably lower than the "decoupling" temperature, at which

$$\rho_{m0}(R_0/R_d)^3 = \rho_{m0}(T_d/T_0)^3 = \sigma T_d^4 / c^2 \qquad (4.51)$$

This leads to

$$T_d = \rho_{m0}c^2 / \sigma T_0^4 = 1.06 \times 10^3 \text{ K} \qquad (4.52)$$

where $\rho_{m0} = 1.97 \times 10^{-31}$ g/cm^3 and $T_0 = 2.796$ K have been used.

In 1948 Alpher and Herman[10] used the relationships for radiation mass density and matter mass density valid beyond the decoupling temperature, to get a prediction of the expected background radiation temperature, which came to be very close to the value later observed by Penzias and Wilson.[11] Their argument may be summarized as follows.

Since

$$\rho_r = \sigma T^4 = \text{constant} \times R^{-4} \qquad (4.53)$$

and

$$\rho_m = \text{constant} \times R^{-3} \qquad (4.54)$$

the quantities for ρ_r and ρ_m at present (0) and at the time of nucleosynthesis (ns) should be related by

$$\rho_{r0} / \rho_{m0}^{4/3} = \rho_{rns} / \rho_{rns}^{4/3} \qquad (4.55)$$

and, therefore

$$T_0 \approx T_{ns}(\rho_{m0} / \rho_{mns})^{1/3} \qquad (4.56)$$

They used $T_{ns} \approx 5 \times 10^8$ K (corresponding, roughly, to $\rho_{ms}/c^2 \approx 1$ g/cm^3), $\rho_{m0} \approx 10^{-30}$ g/cm^3 and $\rho_{mns} \approx 10^{-6}$ g/cm^3, resulting in a value for the present background radiation temperature of $T_0 \approx 5$ K.

Fig. 4.7. Penzias & Wilson.

4.4. The Cosmic Baryon to Photon Ratio

As mentioned before, the ratio of heavy particles or baryons (protons and neutrons) to photons (massless quanta of radiation) in the universe can be easily calculated once the matter density and the radiation density are known. This can be done in the following way:

$$\eta = \frac{n_b}{n_r} = \frac{\rho/m_p}{\sigma T^4/2.8 k_B T} \qquad (4.57)$$

where n_b and n_r are the respective densities of baryons and photons, obtained dividing p_m by the rest mass of a nucleon (m) and $\rho_r = \sigma T^4$ by the average energy of a photon ($\approx 2.8\ k_B T$), respectively. Using

$$\rho_{m0} \approx 1.97 \times 10^{-31} \text{ g/cm}^3, \ m_p = 1.67 \times 10^{-24} \text{ g} \qquad (4.58)$$

$$\rho_{r0} \approx (7.63 \times 10^{-15})(2.796)^4 = 4.66 \times 10^{-13} \text{ erg/cm}^3,$$
$$2.8 k_B T_0 = 1.08 \times 10^{-15} \text{ g} \qquad (4.59)$$

where the Boltzmann constant value $k_B = 1.38 \times 10^{-16}$ erg/K has been substituted, one gets for the present baryon to photon ratio

$$\eta_0 \approx 2.73 \times 10^{-10} \text{ baryons per photon} \tag{4.60}$$

Using a larger value for ρ_{m0}, for instance $\rho_{m0} \approx 10^{-30}$ g/cm^3, one gets $\eta_0 \approx 2.73 \times 10^{-9}$ baryons per photon, which still is a very small number. There are many more photons in the universe at present than protons and neutrons. Since the largest is the number of photons the largest is the entropy[12], this implies a very large entropy for the present universe. But, as mentioned previously, if one assumes throughout the expansion that matter and radiation do not interact with each other, this ratio may have been constant form very early times. Indeed, due to the fact that, according to Eq. (4.43) ρ_m = constant $\times R^3$, we can write the baryon to photon ratio as

$$\eta = \text{const} \times \frac{R^{-3}}{T^3} = \frac{\text{constant}}{(RT)^3} = \text{constant} \tag{4.61}$$

making use of the result in Eq. (4.47).

In words of S.Weinberg,[13] "the occurrence in cosmology of a pure number as small as one part per 1000 million has led some theorists to suppose that this number is really zero-that is, that the universe really contains an equal amount of matter and antimatter. Then the fact that the baryon number per photon appears to be one part in 1000 million would have to be explained by supposing that, at some time before the cosmic temperature dropped below the threshold temperature for nuclear particles, there was a segregation of the universe into different domains, some with a slight excess (a few parts per 1000 million) of matter over antimatter, and other, with a slight excess of antimatter over matter... The trouble with this idea is that nobody has seen signs of appreciable amounts of antimatter anywhere in the universe". Somewhat below, he adds, "Another possibility is that the density of photons (or, more properly of entropy) has not remained proportional to the inverse cube of the size of the universe. This could happen if ... some sort of friction or viscosity could have heated the universe and produced extra photons".

Within the frame work of the Grand Unified Theories[14] a justification of the small value $\eta \approx 10^{-10}$ has been proposed, making a connection between the small baryon antibaryon asymmetry, resulting from the asymmetric decay of the hypothetical massive X-particle (abundantly present at an early high temperature $T_x \approx 10^{28}$ K) ,and the small baryon to photon ratio. According to this proposal

$$\varepsilon = r - \overline{r} = 1 - n_{\overline{b}}/n_b \tag{4.62}$$

where r is the probability of decay of X into two quarks, $\overline{r}$ the probability of decay of X into two antiquarks, n_b the baryon density and $n_{\overline{b}}$ the antibaryon density. Then, if $r \neq \overline{r}$ by a small amount, say 10^{-10}, $\eta = (n_b - n_{\overline{b}})/n_r$ would be also of the order of 10^{-10}.

Conventional wisdom, now and at the time of the writing of the first edition of "The Intelligible Universe" (1993) held that the present matter-dominated epoch was preceded by a radiation-dominated epoch. In my opinion (see comments in Chapter 20) the present epoch was preceded by an epoch in which radiation and matter densities were equal.

Bibliography

1. A. Einstein, "The Meaning of Relativity" (Princeton NJ: Princeton University Press, 1922).
2. S. Weinberg, "Gravitation and Cosmology", pp. 188–209 (New York: Wiley, 1972).
3. Ibid., pp. 470–75.
4. We may note that, instead of Eq. (4.31), the relationship

$$R = \frac{4\pi}{3}G(\rho + 3p)R$$

   is often given in the literature. This equation together with Eq. (4.1) is completely equivalent to Eqs. (4.3) and (4.1) combined.
5. A. Friedmann, Z. Phys. 10, 377 (1922); ibid 21, 326 (1924).
6. J. Silk, "The Big Bang", p. 191 (San Francisco: W.H. Freeman, 1980).
7. Ibid., p. 68.
8. H.A. Brück, G.V. Coyne and M.S. Longair eds. "Astrophysical Cosmology" (Vatican City: Pontificia Academia Scientiarum, 1982).
9. P.H Anderson in "Physics Today", p. 17 Part I, (August 1988).
10. R. Alpher and R. Herman, Nature 162, 774 (1984).

11. A.A. Penzias and R.W. Wilson, Astrophysical J. 142, 419 (1965).

12. S. Weinberg, "The first three minutes", pp. 89–90 (New York: Bantam Books, 1980).

13. Fang Li Zhi and Li Shu Xian, "Creation of the Universe" (Singapore: World Scientific, 1989).

Chapter 5

The Fundamental Physical
Forces in the Universe

5.1. Gravitational, Electromagnetic and Nuclear Forces

There are four[1] fundamental forces in the physical world surrounding us:
the gravitational force associated with mass, the electromagnetic force
associated with electric charge, the strong nuclear force associated with
the stability of atomic nuclei, and the weak nuclear force responsible for
the β-decay of nuclei. Why four, and not five or one? We do not know for
sure. This is a good question which has kept busy theoretical physicists
from Einstein and Heisenberg to our day. We may or may not in the
future find a satisfactory answer to this question, but some progress has
been made recently[2] towards the apparent unification of the
electromagnetic and the weak nuclear force into the so called electroweak force, of which we will speak in a little more detail later. On the
other hand a few years ago considerable stir was produced in the
community of physics by the proposal of a fifth interaction which would
modify Newton's gravitational law at finite distances. This proposal has
not been substantiated by subsequent experimental tests.

The gravitational force, which acts between bodies and particles
having mass, is the most notorious force in the universe. The strength of
this force between two bodies of mass m_1 and m_2, as it is well known, is
given by

$$F_g = Gm_1m_2 / r^2 \qquad (5.1)$$

where $G = 6.67 \times 10^{-8}$ (cgs units) is Newton's gravitational constant and r the distance between the center of mass of m_1 and m_2. Thus the coupling strength for two protons at any distance can be given by

$$\alpha_g = Gm_p^2 / \hbar c = 5.83 \times 10^{-19} \text{ (dimensionless)} \qquad (5.2)$$

Fig. 5.1. Newton.

The interaction can be thought of as being mediated by gravitons, massless particles which play the same role as photons (radiation quanta) in the electromagnetic interaction. Since the particles mediating the interaction are massless, the range of interaction is infinite, i.e. the interaction becomes smaller and smaller with increasing separation between particles but it is still different from zero at very large distance.

The electromagnetic force acts between two electrically charged particles or bodies. A stationary charge gives rise to an electric field, and an oscillating or accelerated charge gives rise to electromagnetic waves, which propagate at the velocity of light through free space or through a material medium. In electrostatic cgs units the electromagnetic force between two static charges q_1 and q_2 is given by

$$F_e = q_1 q_2 / r^2 \qquad (5.3)$$

where the proportionality constant is unity in this case, due to the choice of charge unit, and r is again the distance between the center of charge of q, and q_2. Unlike the gravitational interaction, the electromagnetic one can be positive or negative, depending on the relative sign of the two charges involved. Ordinary matter consists of atoms with equal amounts of positive charge in the nucleus and negative charge in the electron cloud surrounding it. Matter in the form of plasma, like the matter in the sun's interior, contains also equal amounts of free positive charge in protons and negative charge in electrons. The coupling strength for two electrons (repulsive) or two protons (repulsive) or one electron and one proton (attractive) at any distance is given by

$$\alpha = e^2 / \hbar c \qquad (5.4)$$

Fig. 5.2. Faraday.

Thus it is enormously stronger than the gravitational coupling strength for the same particles. Celestial bodies, however, which are made of very large amounts of massive particles, are electrically neutral. The electromagnetic interaction can be thought of as being mediated by massless photons and its range of interaction is therefore infinite, as in the case of the gravitational interaction. It may be noted that, originally, electric forces between electric monopoles (single charges), dipoles (pairs of positive and negative charges separated by a small distance), etc., and magnetic forces between magnetic dipoles (electric current closed loops), etc., where thought to be independent forces, unrelated to one another. The experimental and theoretical work of such great nineteenth century physicists as Oersted, Ampere and Faraday prepared the way to the unification by Maxwell of these two apparently different forces into one single interaction, the electromagnetic interaction. This was done by setting forth the famous four Maxwell equations, which interconnect in an almost symmetrical way E (electrical field) and H (magnetic field) making each one depending of the other. The only asymmetry is that, apparently, there are in nature no magnetic monopoles or, at least, they have not been observed yet. The successful unification by Maxwell of electrical and magnetic forces prompted Einstein in the 1920's to try, unsuccessfully, the unification of electromagnetic and gravitational forces.

Fig. 5.3. Maxwell.

The strong nuclear force acts between nucleons (protons, p, and neutrons, n) to held them together within the atomic nucleus and also between "quarks" (the nucleus elementary constituents, not directly observables but known to exist from high energy scattering experiments) to confine them tightly within the boundaries of a given proton or neutron. The strong force between nucleons (Yukawa theory[3]) may be written as

$$F_s = \frac{g_s^2}{r^2} e^{-\frac{r}{r'}}$$ (5.5)

where g_s plays a role analogous to the mass and the charge in the previous cases, and the exponential factor reflects the fact that these are short range forces which vanish for $r \gg r'$. This interaction, always attractive and nearly the same for the three cases p-p, n-n and n-p, is mediated by massive particles, the pions ($\pi^{\pm}$, π^o), whose mass can be connected to the range (r') by means of the uncertainty principle

$$\Delta p \Delta r \approx \hbar$$ (5.6)

where Δp is uncertainty in momentum of the mediating particle, of the order of the momentum itself, which can be taken as (n\c), and Δr is the uncertainty in position of the same particle, which can be taken as (r'), leading to

$$r' \approx \hbar/m_\pi c^2 \approx 1.4 \times 10^{-13} \text{ cm}$$ (5.7)

This is the scale of size of the nucleus, orders of magnitude smaller than that of the atom. The description, on the other hand, of the strong force acting within individual nucleons is far more complicated. The nucleon constituents are the famous "quarks", and the interaction binding them together is thought to be mediated by "gluons", very heavy particles which produce such a strong attractive force amongst the constituent quarks that the highest available energies in present day accelerators are not sufficient to break down the nucleon into quarks ("strong confinement"). On the other hand the coupling constant when the quarks approach one another at very short distances becomes negligibly small ("asymptotic freedom"). The coupling strength for the strong interaction at the energy scale of distance corresponding to exchange of pions is given by

$$\alpha \equiv g_s^2 / \hbar c \approx 1$$ (5.8)

or about one hundred times stronger than the electromagnetic interaction. In this connection it is worth recalling that there are stable nuclei with up to about one hundred protons (uranium), which means that the electrostatic repulsion corresponding to about one hundred charged protons is more or less compensated by the short range strong nuclear attraction between protons and neutrons within the heaviest stable nuclei.

Finally, the weak nuclear force is responsible, among other things, for radioactive decay (an example is the weak decay of a free neutron into a proton, and electron and an antineutrino). It also participates in such events as the scattering of neutrinos by electrons[4] and protons, and, in general, in interactions between leptons (particles of light mass as the electron, the muon and the neutrino) which often take place with time scales much slower than those characteristic of the strong nuclear interactions. At the energy scale corresponding to the proton mass ($m_p c^2 = 1.5 \times 10^{-3}$ erg $= 0.939$ GeV) the coupling strength of the weak interaction may be given by

$$\alpha_w \approx g_w^2(m_p) / \hbar c = G_F m_p^2 / \hbar c \approx 10^{-5} \qquad (5.9)$$

where g_w plays the role of a weak interaction charge, G_F is the Fermi constant and m_p is the proton mass. This interaction is a very short range one, negligible for $r \gg r''$, being r'' much smaller than the range r' of the strong interaction. The mediating particles, the heavy bosons W^+, W and Z^0, have masses of the order of 80 to 90 GeV/cm (i.e. 80 to 90 times the mass of a proton or neutron) and the same argument used before, relaying on the uncertainty principle, allows an estimate of the range,

$$r'' \approx \hbar / M_{w^\pm} c \approx \hbar / M_{Z^0} c \approx 2 \times 10^{-16} \text{ cm} \qquad (5.10)$$

At these small distances, due to the weakening of the strong force resulting from "asymptotic freedom", the weak nuclear force becomes important[5] in interactions between elementary particles.

Table 5.1 summarizes some prominent features of the four physical interactions discussed above.

Table 5.1. Prominent features of the four physical interactions.

Force	Strength	Range	Mediating Particles
Gravitational	6×10^{-39}	Infinite	Gravitons
Electromagnetic	1/137	Infinite	Photons
Strong	1	10^{-13} cm	Hadrons
Weak	10^{-5}	10^{-16} cm	$W^\pm, Z^0$

Fig. 5.4. Fermi.

The experimental discovery in 1983 at the UA1 detector of CERN Synchrotron facility[6] of the W* and Z^0 particles, the heavy bosons that mediate weak interactions, clinched the case also for the theoretical unification of the electromagnetic and the weak interactions. It is well known that Quantum Electro Dynamics (QED) is a gauge theory, i.e., a theory in which the dynamic equations describing the physical system in question are unchanged by certain (gauge-symmetry) transformations of the potentials appearing in them. In QED the photon is the boson-like particle associated with changes in the fields produced by electrically charged particles. In the unified electroweak theory, which presents a larger class of gauge symmetries, the variables describing particles of different electric charge (the electron and the neutrino, for example) can be interchanged without changing the form of the equations. In addition to the massless photon, three new particles, the three massive vector bosons W^+, W'', and Z^0 are required by the theory. A v_μ (μ neutrino) - p (proton) scattering event is mediated by a Z^0 particle exactly in the same way as a e (electron) -p (proton) scattering event is mediated by a photon (see diagrams in Fig. 5.5).

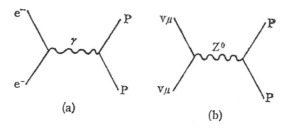

Fig. 5.5. (a) electron-proton scattering mediated by a photon (γ); (b) neutrino-proton scattering mediated by a neutral vector boson (Z^0).

5.2. Conservation Laws

The existence of conservation laws plays a fundamental role in all branches of Physics. These laws cannot be taken as arbitrary restrictions imposed on physical processes by the observer. On the contrary, they are general statements about hard facts of nature. The enunciation of the law of conservation of matter by A.L. Lavoisier (1743–1794) was the foundation upon which rested future developments in the understanding of physico-chemical processes. At the same times it prepared the way for the ordered classification of the atomic elements into the periodic table in the late nineteenth century. The realization by J.R. Mayer (1814–1878), and almost simultaneously by other contemporary scientists, of the law of conservation of energy was instrumental in producing a fruitful unity in the whole realm of physical science, from biochemistry, to mechanics, to thermodynamics and electromagnetism. This opened the way to still higher achievements in the years to come. The more general and deeper law of mass-energy conservation, established by A. Einstein (1879–1955) on the grounds of his Theory of Relativity, allowed the subsequent understanding of subatomic processes, governed by the strong and weak interactions, such as nuclear fission and fusion. An interesting example of the fruitfulness of conservation laws is the one which led to the discovery of the neutrino. When Enrico Fermi was investigating in 1934 the spontaneous decay of the unstable neutron into protons and electrons he observed that electric charge was conserved in the process of disintegration, but energy and momentum were not, because the combined energy of proton and electron was less than that of the original neutron. Therefore Wolfgang Pauli postulated the existence of a new particle, the neutrino, with no charge and zero or very small rest mass, which could pick up the extra energy. The decay process would then be described by

$$n \rightarrow p + e + v \qquad (5.11)$$

This was a very bold assumption because as shown by later work, a neutrino interacts very weakly with ordinary matter and it is therefore very difficult to observe (it is estimated that a neutrino traveling straight through a solid lead tube from Alfa Centaury, four light years away from us, to the Earth, would have an almost neglible probability of being scattered by the closely packed lead atoms along its path). But otherwise the law of mass-energy conservation would have been violated, and not only this, but also the law of conservation of spin (internal angular momentum of a particle, which is quantized). This law could be restored if the neutrino (we know now that it is really an antineutrino, rather than a neutrino) had also a specific spin. In 1956 Clyde L. Cowan and F. Reines were able to demonstrate experimentally,[7] using the massive flux of antineutrinos coming from a nuclear reactor (10^{18} neutrinos/second) and a large number of detectors surrounding it, that the reaction

$$v_e + p \rightarrow n + e+ \qquad (5.12)$$

was in fact taking place, thus demonstrating the existence of the electron antineutrino as a real new particle.

Every continuous symmetry entails the existence of a constant of motion which is a conserved quantity. Systems that are invariant under time translation conserve energy; systems invariant under space translation conserve momentum; systems invariant under rotation conserve spin (angular momentum). There are a number of other, more special, conservation laws, discovered by particle physicists, which include, in addition to energy, momentum, spin, and electric charge, the conservation of other less familiar quantities,[8] such as muon, lepton and baryon number, charge conjugation, parity, strangeness, charm and isotopic spin. Not all interactions conserve all these quantities. For example isotopic spin, which is related to the number of different particles with charge +, - or 0 in a given family of alike mass particles, is conserved only by the strong interaction.

The introduction of these perfect (the familiar ones) and imperfect (the less familiar) conservation laws is useful for the classification of elementary particles in related families, and makes it possible to correlate many properties within the same family of particles, f. i. their masses. These less familiar conservation laws are related to underlying symmetries whose physical nature is not clear to us for the time being.

Cosmological models which take lightly the law of conservation of mass-energy, like the former steady-state model or some of the more recent "inflactionary" models, must be considered with great caution. Once rabbits are allowed to get out of a hat, entire galaxies may be soon allowed to come out too. In this connection it may be remembered that Heisenberg's uncertainty principle restricts the value of the minimum observable energy in a given microscopic process in inverse proportion to the uncertainty in time of the same process. It is well known that this restriction becomes irrelevant[9] for isolated macroscopic systems, and the universe is such an isolated and macroscopic system, by definition.

5.3. Elementary Particles

The investigation of natural radioactive process, as well as fission and fusion processes, shows that the main elementary constituents of matter are the proton (stable), the neutron (stable within the nucleus but unstable outside it, with a lifetime of about ten minutes) and the electron (stable). With the advent of high energy accelerators and sophisticated detectors physicists all over the world begun to discover an increasing amount of new particles, most of them extremely short lived, whose existence was deduced from peaks ("resonances") in measurements of the cross section (proportional to the occurrence probability) for a given event as a function of the incident energy of the bombarding particle. Matt Roos[10] published in the early sixties the first catalog of the then known "particles" (17) and "resonances" (24) in "Reviews of Modern Physics". The latest version that he published in 1976 was a very long list making up 245 pages (with a supplement of another thirty pages) containing literally hundreds of "elementary" particles and resonances. It was clearly necessary to introduce some unity in such a variegated multiplicity. To do a detailed discussion of the way in which theoretical physicists have been able to put order and simplicity in the field of elementary particles is beyond the scope of this book.

However, without entering into the details of how this remarkable achievement took place we may carry out a little further the parallelism with the classification of atomic elements into the periodic table. The periodic table of the chemical elements was arranged ordering those elements according to increasing atomic weight and breaking the series in a set of rows in such a way that elements with the same chemical valence (and therefore with the capability to form alike chemical compounds with other specific elements) would fall on top of each other. Thus we have H, Li, Na, K, etc. on the first column, Be, Mg, Ca, etc. in the next column, and so forth. So we may say that we are classifying the atoms by increasing weight down the vertical axis and by increasing chemical valence to the right of the horizontal axis. The symmetry of the arrangement of stable elements in nature is not perfect (there are rare Earth elements to complicate the picture) but it was good enough to allow predictions about where as yet undiscovered elements should fit in the table, and on what their mass and chemical properties should be. Later, Quantum Mechanics provided a satisfactory justification of this classification scheme.

In 1961 Murray Gell-Mann, and, independently, Yuval Ne'eman, found a way to group orderly the reduced number of particles and/or resonances which disintegrated with similar end products, representing $(B + S)$ in the horizontal axis versus (I_z) in the vertical axis. Here B, S, and I_z are characteristic quantum numbers of the particle or resonance, deduced from their decay products. B is the "baryon" number or the number of baryons (protons + neutrons) at the end of the disintegration process. S is the "strangeness" quantum number, or the number of "strange" particles, relatively long lived resonances or particles, 10^{-10} seconds, in contrast to the "normal", short lived resonances or particles, 10^{-23} seconds, which appear in the decay products. Finally I_z is the projection along the z axis of the so called "isospin" of the particle, defined, in an abstract space so as to describe the numbers of varieties of charge, f.i. +1, 0, -1, for that particle, in analogy with the quantized values of ordinary spin in angular momentum space. The number of different possible states for a system with spin I is given by $N = 2I + 1$. The number of charged varieties of the nucleon is two $(Q(p) = +1$, $Q(n) = 0)$, therefore $I = (N-1)/2 = 1/2$, and $I_z = +1/2, -1/2$. The possible values of I_z for other particles can be found from an experimental knowledge of N. For instance for the pion (π^+, π^-, π^0), $N = 3$, and therefore $I_z = +1, 0, -1$, and so on.

The procedure devised by Gell-Mann and Ne'eman allowed to classify by their (B + S) and I_z quantum numbers many of the particles and resonances in Roos list, forming octects (with one particle in each vertex of an hexagon and two occupying the centers, see Fig. 5.6), and it received the exoting designation of "eightfold way" (after a text attributed to Buda). Later, other families of particles, like the spin 3/2 baryons, were classified in the same way forming a decuplet (with four successive rows of, respectively, four three, two and one particles, producing a triangle, see Fig. 5.7). Both hexagons and triangles have threefold rotational symmetry, which in group theory is designated as SU(3). Gell-Mann and Okubo,[11] on the ground of the information provided by the latter incomplete decuplet, predicted the mass of the Ω^- particle, soon after discovered at the bubble chamber of Brookhaven National Laboratory (and shown to posses the predicted mass). This clinched the case for SU(3) as a successful theory for the understanding of elementary particles, and led Gell-Mann to postulate the existence of "quarks" (fractionally charged particles), which in various combinations could form the known mesons and baryons.

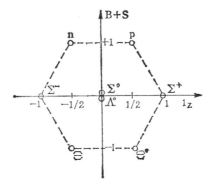

Fig. 5.6. The even parity spin 1/2 baryon octect containing the nucleons.

Up to the present time three "generations" of "quarks" have been predicted, and each family, containing three varieties or "colors" of quarks (R = red, G = green, B = blue) is associated with a lepton and a neutrino. Each generation contains two varieties of quarks: first (d = down, u = up), second (s = strange, c = charm) and third (b = bottom, t = top). Except the t quark, all others have been indirectly detected in high energy events.

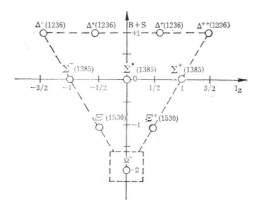

Fig. 5.7. The even parity spin 3/2 baryon decuplet containing the Ω^- whose mass was successfully predicted from the known masses (in GeV in parentheses) of the other particles in the decuplet.

Table 5.2 gives a list[12] of the known elementary constituents of matter.

Table 5.2. Known elementary constituent of matter.

	Leptons	Quarks
1st generation	e^- ν_e	$d_R d_G d_B$ $u_R u_G u_B$
2nd generation	μ^- ν_μ	$s_R s_G s_B$ $c_R c_G c_B$
3rd generation	τ^- ν_τ	$b_R b_G b_B$ $t_R t_G t_B$

For instance, neutron decay, $n \rightarrow p + e + \bar{\nu}_e$ put in terms of quarks ($n = ddu$, $p = uud$) is given by

$$
\begin{aligned}
d &\rightarrow u + W^- \rightarrow u + e^- + \bar{\nu}_e \\
d &\rightarrow \qquad\quad \rightarrow d \\
u &\rightarrow \qquad\quad \rightarrow u \\
\hline
(ddu) &\rightarrow \qquad\quad \rightarrow (uud) + e^- + \bar{\nu}_e
\end{aligned}
$$

The number of particles in Table 5.2, twenty four, plus that of the respective antiparticles, gives a total of 48 elementary particles. One of the tasks of a unified theory is that of representing precisely this particle content.

5.4. Universal Constants

The concept of universal constant may be traced back to Planck's theoretical work on blackbody radiation.[13] His quantum theory, which, in a single stroke, was successfully to account for

(a) the spectral distribution from the lowest to the highest frequency in the blackbody radiation[14] by means of

$$W_T(\omega)d\omega = \frac{\hbar}{\pi^2} \frac{\omega^3}{c^3} \frac{1}{e^{\hbar\omega/k_B T}} d\omega \qquad (5.13)$$

Fig. 5.8. Planck.

where W_T is the thermal radiation energy density per unit volume and per unit frequency interval, the frequency and T the absolute temperature,

(b) the Wienn displacement law, according to which the maximum of the emitted radiation for a given equilibrium temperature T occurs at a certain frequency ω_{max} such that it moves up with T as

$$\hbar\omega_{max} \approx 2.8 k_B T \tag{5.14}$$

(c) the Stefan-Boltzmann law, giving the total amount of emitted radiation per unit volume in the whole frequency range from zero to infinity,

$$\int_0^\infty W_T(\omega)d\omega = \left[\frac{\pi^2}{15}(\hbar c)^{-3}k_B^4\right]T^4 \equiv \sigma T^4 \tag{5.15}$$

allowed Planck to determine the two universal physical constants[15]

$$\hbar = h/2\pi = 1.05 \times 10^{-27} \text{ erg} \cdot \text{s} \tag{5.16}$$

$$k_B = 1.38 \times 10^{-16} \text{erg/K} \tag{5.17}$$

And to relate them to the other universal physical constant

$$c \approx 3 \times 10^{10} \text{ cm/s} \tag{5.18}$$

the velocity of light in free space, appearing in Eqs. (5.11) to (5.13). The founder of quantum theory was quick to point out that the introduction of his universal constant made possible the derivation of units of mass, length, time and temperature which are "independent of specific bodies and substances, and necessarily keep their meaning for all times and for all cultures, even for extraterrestrial and extrahuman creatures, and which can be designated as 'natural units'"[16]

The itinerary of Planck's approach through science to the absolute is best summarized in his own words: "What has led me to science and made me since youth enthusiastic for it is the not at all obvious fact that the laws of our thoughts coincide with the regularity of the flow of impressions which we receive from the external world, [and] that it is therefore possible for man to reach conclusions through pure speculation about those regularities. Here it is of essential significance that the external world represents something independent of us, something absolute which we confront, and the search for the laws valid for this absolute appeared to me the most beautiful scientific task in life."[17]

Let us see how one can get, in the spirit of Planck's reasoning, natural units of mass, length and time, and how are they related to the mass, size and age of the universe as far as we know them. Consider a single high energy photon in which the gravitational pressure associated with its mass is compensated by its own radiation pressure. In this case we can equate the gravitational and the radiation energy density by

$$\left[G \frac{\left(\rho_r / c^2\right)^2 \left(\frac{4\pi}{3} \cdot r_0^3\right)^2}{r_0} \right] \Bigg/ \left(\frac{4\pi}{3} \cdot r_0^3\right) \approx \rho_r \tag{5.19}$$

where G is Newton's gravitational constant, the energy density associated with the photon, and r the radius of the volume occupied by this photon. Taking into account the Stefan-Boltzmann law, Eq. (5.15), which states that

$$\rho_r = \sigma T^4 \tag{5.20}$$

substituting it into Eq. (5.19) and rearranging terms, we get

$$T^4 r_0^2 \approx c^2 / (4\pi/3) G\sigma = \text{constant} \tag{5.21}$$

On the other hand the fact that we are considering a single photon implies

$$n_r V = \left(\frac{\sigma T^4}{2.8 k_B T} \right) \frac{4\pi}{3} r_0^3 = 1 \tag{5.22}$$

and, therefore

$$T^3 r_0^3 = 2.8 k_B / (4\pi/3)\sigma = \text{constant} \tag{5.23}$$

Combining Eqs. (5.21) and (5.23) we get

$$2.8 k_B T / r_0 \approx c^4 / G \tag{5.24}$$

We can now make use of the relativistic equation connecting mass and energy, to write down

$$M_p = 2.8 k_B T / c^2 \tag{5.25}$$

and of the quantum uncertainty relationship connecting $\Delta r = r_0$ and $\Delta p = M_p c$ to get

$$r_p \approx \hbar / M_p c \tag{5.26}$$

Substituting Eqs. (5.25) and (5.26) into Eq. (5.24) we get

$$\frac{M_p c^2}{\hbar / M_p c} \approx \frac{c^4}{G}, \text{i.e. } M_p \approx \left(\frac{\hbar c}{G} \right)^2 \approx 2.1 \times 10^{-5} \text{ g} \tag{5.27}$$

which is Planck's natural unit of mass, or the mass of a Planck monopole. Planck's length can be obtained from Eqs. (5.26) and (5.27) as

$$L_p = r_0 = \left(\hbar G / c^3 \right)^{1/2} \approx 1.6 \times 10^{-33} \text{ cm} \tag{5.28}$$

and Planck's time, directly related to Planck's length, as

$$\tau_p = L_p / c = \left(\hbar G / c^5 \right)^{1/2} \approx 5.4 \times 10^{-44} \text{ s} \tag{5.29}$$

These are indeed a short length and a short time, much shorter than anything that can be actually measured in the laboratory with the most precise instrumentation. On the other hand Planck's mass is fairly sizable, of the order of 10^{19} larger than the proton mass.

We may give cosmic quantities such as the mass, radial size and age of the observable universe in terms of these units. This is done in the accompanying table (Table 5.3).

Table 5.3. Cosmic quantities.

	Cosmic quantity		Natural unit	Ratio
Mass	$M_U \approx 3.8 \times 10^{55}$ g		$M_p \approx 2.1 \times 10^{-5}$ g	1.8×10^{60}
Length	$R_0 \approx 1.6 \times 10^{28}$ cm		$L_p \approx 1.6 \times 10^{-33}$ cm	1.0×10^{61}
Time	$\tau_0 \approx 5.5 \times 10^{17}$ s $\simeq 1.75 \times 10^{10}$ year		$\tau_p \approx 5.4 \times 10^{-44}$ s	1.0×10^{61}

It is not surprising that (R_0/L_p) and (τ_0/τ_p) give the same large ratio, since $R_0 \approx \tau c_0$, but it is somewhat surprising that this ratio comes so close to that of (M_U/M_p). It must be noted, however, that the later is a true constant ratio, if energy is conserved, as it should, through the time evolution of the universe, while the former are continuously changing with time as the universe expands and gets older.

There is also a Planck's temperature, associated with Planck's mass, which is given by

$$T_p \approx M_p c^2 / 2.8k_B \approx (\hbar c^5 / G)^{1/2} / 2.8k_B \approx 5 \times 10^{31} \text{ K} \qquad (5.30)$$

This is an enormous temperature, much larger than any temperature in the interior of a star, for instance, and, if the Big Bang concept gives a correct description of the time evolution of the universe, it must have been the characteristic temperature of the cosmos at a very short time after the beginning, of the order of Planck's time.

5.5. Understanding the Universe, and Open-Ended Process

In view of what been succinctly summarized in this and previous chapters it seem reasonable to say that the physical universe is indeed intelligible. This does not mean, of course, that we understand everything in the physical universe or even that we will ever be able to understand everything. Our observation capabilities are finite, and so are out capabilities of systematization, as attested by the famous Gödel theorems.[18] A corollary in these theorems holds that in any mathematical system which is consistent (either "de facto" or assumed to be so) it is impossible to find the proof of consistency within it. But this does imply only that the wonderful process of understanding the universe is for us an open-ended process and will remain so in the future, no matter how large and how good computers are available.

Some among us may not like it at all, but only God understands everything. However, this does not detract a bit from the beautiful intellectual adventure of man understanding the universe. The science of physics, through the work of a handful of geniuses and a legion of obscure investigators, has lead us to a most unified and most coherent vision of the universe, in which hard experimental facts and elegant mathematical theories are interconnected in an admirable way. Never mind that often these "hard" facts and theories are almost unavoidably mixed with "soft" speculations. In the long run, and precisely because the Creator of both men and universe has endowed us with an intellect, however fallible, able to detect the true ring of truth, we may be confident that the hard facts and the good theories will be incorporated in a future better vision of the cosmos, and the soft speculations will be discarded in due time.

Bibliography

1. L. Dolan, "Unified field theories", in "Encyclopedia of Physical Science and Technology", Vol. 14, pp. 220–30 (New York: Academic Press, 1987).
2. P. Langacker and A.K. Mann, p. 22 "Phys. Today" (December, 1989).
3. R. Eisberg and R. Resnick, "Quantum Physics of Atoms, Molecules, Solids, Nuclei and Particles", pp. 685–92 (New York: John Wiley and Sons, 1974).
4. L.A. Ahrens et al. (E734 experiment) Phys. Rev. Lett. 54, 18 (1985); K. Abe et al., Phys. Rev. Lett. 62, 1709 (1989).
5. D.J. Gross, p. 39, "Phys. Today" (January 1987).
6. G. Arnison et al. (UAI collaboration), Phys. Lett. B 166, 484 (1986); C. Albajar et al., Z. Phys. C 44, 15 (1989); R. Ausari et al. (UA2 collaboration), Phys. Lett. B 186, 440 (1987).

7. J.S. Trefil, "From atoms to quarks", Chap. 13 (New York: Ch. Schribner's Sons, 1980).

8. Ibid., Chap. 8.

9. S.L. Jaki, "God and the Cosmologists", pp. 129–135 (Washington D.C.: Gateway, 1989).

10. J.S. Trefil, "From atoms to quarks", Chap. 7 (New York: Ch. Schribner's Sons, 1980).

11. R. Eisberg and R. Resnick, "Quantum Physics of Atoms, Molecules, Solids, Nuclei and Particles", p. 708.

12. Fang Li Zhi and Li Shu Xian, "Creation of the Universe", p. 23 (Singapore: World Scientific, 1989).

13. R. Loudon, "The Quantum Theory of Light", 2nd. ed., Chap. 1 (Oxford: Claredon Press, 1983).

14. Planck used in fact h instead of ℏ, introduced later by Dirac, but this is of no consequence for our purpose.

15. Max Planck, "Physikalische Abhandlungen und Vortráge", 1:666 (Braunschweig: Friedr. Vieweg & Sohn, 1958). See also "Planck's Original Papers in Quantum Mechanics", edited by H. Kangro (London: Taylor and Francis, 1972).

16. Max Planck, "Wissenschaftliche Selbstbiographie" (1948), p. 374, in "Physikalische Abhandlungen", 3:374 (Braunschweig: Friedr. Vieweg & Sohn, 1958). See also "Planck's Original Papers in Quantum Mechanics", edited by H. Kangro (London: Taylor and Francis, 1972).

17. S.L. Jaki, "Brain, Mind and Computers", pp. 214–222, paperback edition (Gateway: South Bend, Indiana, 1978), and references therein.

Chapter 6

Cosmology and Transcendence

6.1. Towards the Confines of the Universe

As illustrated in the previous chapters, one of the chief achievements of our century has been to give, for the first time in history, a coherent quantitative (even if tentative) description of the entire physical universe. Physicist and astronomers have been amassing a hugh amount of new facts, and reasonable interconnections amongst these facts, which point out with increasing evidence to the conclusion that our Universe was created some fifteen or twenty thousand million years ago in a primordial event of incredible energy known as the "Big Bang". According to Prof. Jastrow,[1] founder and director of the NASA Goddard Institute, these discoveries are truly fascinating "partly because of their religious implications and partly because of the peculiar reactions of my (scientist) colleagues before them."

Jastrow, according to whom scientists did not expect to find evidence for an abrupt beginning, summarizes succinctly the story of these remarkable discoveries. In 1913, Slipher, which was looking for solar systems in formation, discovers that the nebula identified as Andromeda is an "island universe", similar to our galaxy, which is receding from it at a fantastic velocity, as indicated by the Doppler shift of the emitted light. Later, Hubble and Humason extend their observations to more and more distant galaxies and confirm that the recession of galaxies is universal and that the speed of this recession of the galaxies increases approximately in proportion with their distance to ours. Some years later, in 1917, Einstein applies to the cosmos as a whole his new theory of general relativity. Because he expects a static cosmos, he includes at first in his equations a so called cosmological term, which opposes the universal attraction of the gravitational force. This cosmological term was later considered by Einstein "the greatest blunder of his life", but present day cosmologists, refusing to forget it altogether, have invoked it again, in connection with "inflationary" cosmological models. Very soon after Einstein's equations were set forth, the dutch astronomer de Sitter finds special solutions which predict an expanding universe, and, a few years later, the Russian mathematician Alexander Friedmann finds a complete set of solutions (see Chapter 4) which includes one ever expanding solution ($k < 0$), one "marginally" expanding solution ($k = 0$), and one oscillating solution ($k > 0$). By that time news begin to spread out in the scientific community that the observations of Hubble and Humason made with the most potent telescopes then available, seem to support the idea of a non-static, expanding universe.

Fig. 6.1. Slipher.

Einstein travels to visit Hubble in the United States and he appears to overcome his skepticism about the expansion. While in America, he meets the Belgian priest George Lamaître, who had worked out a general and complete set of solutions to Einstein cosmological equations, both with zero and non-zero cosmological term, and had shown convincingly that no static solution was stable. His work had deserved the support of the famous British physicist and mathematician Sir Arthur Eddington, who was instrumental in giving wide publicity lo Lemaître's work.

Fig. 6.2. Einstein & Lemaître.

The Second World War and the events preceding it in Germany distracted the attention of many prominent european and american physicists away from cosmology and often to matters related to the war effort. Some years after the end of the war, as mentioned in previous chapters, G. Gamow's young collaborators R. Alpher and R. Hermán predicted the existence of a remnant background radiation provenant from the "primeval fireball" of incredibly dense energy and elementary particles, out of which nuclei and atoms, galaxies and stars would later be formed. It is interesting to note that they were working on the ground of the "Big Bang" model, somehow implicit in the expanding solutions of Einstein cosmological equations and explicitly pointed out by

Lamaître when he spoke of a "primordial egg" of extremely high mass and energy out of which the material universe evolved. By the late 40's and early 50's there was a rival theory of the "Big Bang" for the origin of the cosmos, the so called "steady state" theory, in which the universe remained more or less the same as time went on, but continuous creation out of nothing had to be assumed to compensate for the expansion. This non-energy-conserving theory,[2] proposed by Bondi, Gold and Hoyle, somewhat surprisingly, enjoyed considerable favour in wide scientific circles, until the experimental confirmation of the background thermal radiation by the American radioastronomers A.A. Pencias and R. Wilson, in 1965, difficult to explain by the "steady state" theory, inclined the balance of the scientific consensus in favour of the "Big Bang" alternative. In the last decade refined experimental observations have confirmed the blackbody and isotropic character of this radiation, and astronomical observations by powerful telescopes in the whole range of radiation have attested to the universal expansion up to the limits of the observable universe.

Fig. 6.3. Alpher & Herman.

For the "philosophers" of the Enlightenment, as well as for most nineteenth century scientists following their steps, the universe was infinite and eternal, and matter continuous and indefinitely subdivisible, in contrast with the well known facts of science today. The best available experimental data seem to be also in contradiction with the ancient concept of the universe[3] as subject to the iron law of "eternal returns". As illustrated by the story of the modern development of scientific cosmology, the conception of the cosmos which it seems to support, often against the hopes of those who have contributed decisively to shape it, seems to be more in consonance with the old biblical conception of a laterally created universe than with the ancient conceptions of past pre-christian cultures. In words of Prof. Jastrow: "For the scientist who has lived by his faith in the power of reason, the story ends like a bad dream. He has scaled the mountains of ignorance; he is about to conquer the highest peak; as he pulls himself over the final rock, he is greeted by a band of theologians who have been sitting there for centuries."

6.2. Observable Data and Big Bang Model

The Big Bang model is based upon a number of well established observational facts, which can be summarized as follows:

6.2.1. *Approximately isotropic distribution of galaxies in space*

From our point of observation (our Earth) the number of observable galaxies aside from our Milky Way, seems to be more or less the same, no matter the direction at which we point out our telescope. Recent investigations of the variation of the density of galaxies with distance to ours, reported the existence of large local concentrations of galaxies (the so called "great attractor" and the "great wall") as well as large voids in cosmic space. These reports have not been supported by close scrutiny. At a sufficiently large scale the concentration of galaxies can be considered roughly homogeneous and isotropic. This means that, observing from another distant point, the universe would look also quasi-homogeneous and quasi-isotropic in galaxy content.

It was mentioned before (Chap. 3) that the fact of a dark sky at night would be contradictory with an infinite universe (Olber's paradox). From this fact we are allowed to conclude that the observable universe is finite. (This is equivalent to say that the farthest away galaxies are escaping from us at nearly the speed of light, but no more, otherwise they would not be observable).

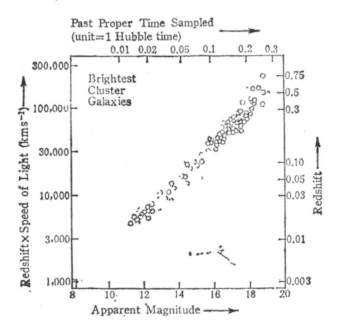

Fig. 6.4. Velocity (redshift) vs. apparent magnitude (distance) for a large representative, sample of galaxies. The upper points, corresponding to brightest cluster galaxies, have the largest uncertainties.

6.2.2. *Universal recession of the galaxies*

The observations of Hubble and Humason, abundantly confirmed by further and more precise observations, lead to the enunciation of Hubble's law:

$$H_0 = \dot{R}_0 / R_0 \approx (75 \pm 25) \text{ km/s} \cdot \text{Mpc} \tag{6.1}$$

where R_0 is the recession velocity of a galaxy (determined from the Doppler shift of its emitted light) and R_0 the distance to us (determined by various methods, f.i., by the apparent luminosity of the brightest stars of the galaxy, assumed to be approximately the same for all galaxies). The Megaparsec, as noted previously, is an astronomical unit of distance which is given by

$$1 \text{ Mpc} \approx 3.26 \times 10^6 \text{ light years} \qquad (6.2)$$

This well established fact of the universal expansion allows us, conceptually, to consider the process in reverse and reach back an instant at which all mass and energy of the universe would appear concentrated in a single point. This occurs at the instant of the Big Bang, considered for some time as an unsatisfactory assumption by many, and generally accepted now by the scientific community as standard reference model to describe the time evolution of the universe.

The above value of the Hubble constant allows (see Chap. 4) an estimate of the time elapsed since the Big Bang event, which would be roughly

$$t_0 \approx 1/H_0 \approx 4.2 \times 10^{17} \text{ sec} \approx 1.3 \times 10^{10} \text{ years}, \qquad (6.3)$$

or somewhere between ten and twenty thousand million years, which is in acceptable agreement with other independent estimates[5] based upon the ages of globular clusters, the age of the oldest stars in our own galaxy and radioactive dating. Within our own solar system (and we must remember that the sun is a second or third generation star) the oldest rocks, as well as lunar samples brought to Earth by american cosmonauts, are about 4 to 6 thousand million years old.

6.2.3. *Relative abundance of ^{4}He and other primordial light elements*

We know that the intermediate and heavy nuclei of the periodic table of the chemical elements are synthesized in the core of stars, under extreme conditions of high temperature and pressure, from lighter nuclei, starting with hydrogen nuclei or protons. The variation in relative abundance of helium (^{4}He in particular) and other few very light elements, on the other hand, seems to be more[6] or less constant throughout a given galaxy and more or less the same for different galaxies. This relative abundance is also by far too large to have been produced by the limited number of successive star-life cycles allowed by the present age of the universe. It suggests a "primordial" origin for ^{4}He and the other very light elements, in contrast to what happens with the intermediate and heavy elements. By "primordial" we mean of early cosmic origin (unrelated to fusion within stars) or having taken place at an early time in the cosmic expansion. At this early time the prevalent conditions of very high density and temperature should have been capable of producing the synthesis of deuterium nuclei (each with a proton and a neutron), and subsequently the synthesis of ^{4}He nuclei (with two protons and two neutrons).

The characteristic temperature for deuterium formation is $T_D \approx 9.3 \times 10^9$ K. At $T > T_D$ the thermal background radiation is sufficient to break down deuterons into its two components, but at $T < T_D$ the background radiation does not have enough energy to break down deuterons, and these become stable. The characteristic temperature for ^{4}He formation, on the other hand, is $T_{4He} \approx 4 \times 10^9$ K. It may be assumed that at $T > T_D$ (very early into the universe expansion) the number of free protons and free neutrons was the same. But the free neutron is unstable[7] (it has a half life of about 700 s) and this implies that if the cooling rate of the universe between T_D and T_{4He} is rapid enough, compared with the neutron half-life, all available neutrons could have been trapped into ^{4}He nuclei, while if this cooling rate is too slow, almost no neutron could have been trapped. The first possibility leads to a universe made up predominantly of helium (incapable of supporting life as we know it), the second to a universe in which helium is almost absent (contrary to what is observed). We know experimentally that the universe contains at present about three times more hydrogen than helium. This is possible only if the cooling rate of the universe is properly tuned to the disintegration rate of the neutron to produce that specific proportion of hydrogen and helium.

Einstein's cosmological equations and the Big Bang model provide a connection between the present ($T_0 \approx 3$ K) expansion rate and the "primordial" ($T_D \approx 9.3 \times 10^9$ to $T_{4He} \approx 4 \times 10^9$ K) expansion rate in such a way that the present helium abundance, given by

$$Y = (^4\text{He mass})/(\text{H mass}) \approx 0.25 \qquad (6.4)$$

is reasonably well justified in terms of the half-life of the free neutron.

6.2.4. *Cosmic background radiation*

As mentioned in previous chapters (Chaps. 1 and 4), in 1948 R.A. Alpher and R. Herman, G. Gamow's collaborators, predicted[8] a background radiation of about 5 K, as a remnant from the very hot state of the early universe, which was confirmed in 1965 experimentally by A.A. Penzias and R. Wilson.

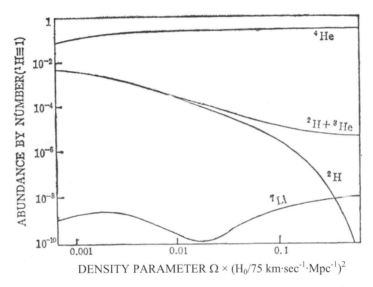

Fig. 6.5. Relative abundance of primordial light elements as a function of Ω_0 (ratio of density t_0 critical density) from cosmic nucleosynthesis. (The central band indicates the range compatible with observations).

These American radioastronomers were looking for something else, and, after confirming that the observed isotropic 3 K background radiation was not instrumental noise, they published[9] their finding in the "Astrophysical Journal" with no further comment.

The alternative model competing with the Big Bang model at that time, the steady-state (continuous creation) model of T. Gold, H. Bondi and F. Hoyle, was unable to give any reasonable explanation for this finding, and this fact contributed considerably to tilt the balance within the scientific community in favor of the Big Bang model. Significant facts

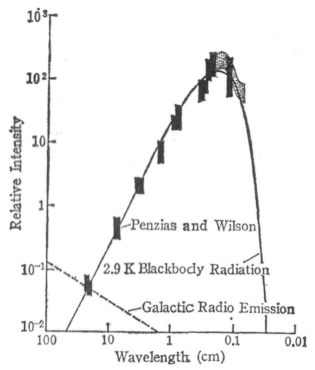

Fig. 6.6. Set of Background radiation data at number of radio microwave, an far-infrared wavelengths (vertical rectangles) showing good fit to theoretical blackbody radiation spectrum at about 2.9 K.

about this background radiation, later confirmed with increasingly precise experiments in the whole range of wavelengths are its typically blackbody spectral distribution and its extreme[10] isotropy (one part in 10^4), after discounting a small anisotropy which was well explained by the relative motion of our Earth, following the sun in its way around the center of the galaxy. These facts could be reasonably taken as pointers to the truly cosmic origin of the background radiation, and as an indication of an early time before stars and galaxies were formed, in which thermal equilibrium prevailed throughout the hot expanding universe.

The experimental observations outlined above give a reasonably firm support to the (once considered wild) hypothesis if a primordial Big Bang event. Of course, many important questions remain open, like the origin and formation of galaxies, the high specific entropy of the universe, the matter-antimatter asymmetry and the specific value of the cosmic space curvature. However, as stated by S.Weinberg, Physics Nobel Prize winner in 1978, in his well known book "The first three minutes",[11] "the great thing is not to be free of theoretical prejudices, but to have the right theoretical prejudices". What has been gained up to now gives us some right to think that we are in the right track, and that the standard model of the Big Bang is a valid reference frame for the study of the cosmos.

6.3. Implications of Contemporary Cosmology

Before briefly analyzing some of the implications of contemporary cosmology it is convenient to remind ourselves that scientific explanations are not, by its own nature, exhaustive, and that they are always exposed to the fire-proof of new observational data.

Nevertheless, substantial progress has been made in our century in the understanding of the physical universe, and this should be seen as having important metaphysical implications. Any man, according to Einstein,[12] being a "two-legged animal is the facto...a metaphysics". It is not science's role to spell out these implications, but science may be, and should be, immensely helpful in providing the ground to make the subsequent metaphysical steps. "Do not be afraid of being freethinkers. If you think strong enough, you will be forced by science to the belief in God, which is the foundation of all religion. You will find science not antagonist but helpful to religion",[13] said Lord Kelvin. He, of course, was a convinced believer, but no less a great contemporary physicist than R.P. Feynman, not a believer by his own admission, acknowledged that "many scientists do believe in both science and God, — the God of revelation — in a perfectly consistent way".[14] Another Nobel laureate, Sir William Bragg, pronounced these eloquent words: "From religion comes Man's purpose; from science his power to achieve it. Sometimes people ask if religion and science are not opposed to one another. They are: in the sense that the thumb and fingers of my hand are opposed to one another. It is an opposition by means of which anything can be grasped".[15]

If the total mass of the observable universe is finite, as already pointed out by Einstein,[16] if its spatial extent is finite, as indicated by astronomical observation,[17] and if its age is also finite, as implied by the Big Bang concept and supported by the available experimental information, then the universe, as we know it, is contingent and created. LamaTtre, Eddington, Jeans, Whitteker and von Weizsáker were among the first to use the term "creation" within the context of physical cosmology, even if it should be recognized that its meaning is not the same for all those that use this word. And there is no "Creation" without a "Creator". The only problem is that, notwithstanding the efforts of such distinguished physicists as Prof. S. Hawking, the act of creation (true creation out of nothing and in time), is outside the realm of physics, among other things because the "nothing" would be physically unobservable.

The current interest in Big Bang cosmology and its implications suggests a quick look back to what happened by the middle of the previous century, when the second principle of thermodynamics was rigorously established by R. Clausius. This principle holds that in any isolated physical system (and such is the universe, by definition) its entropy, or degree of disorder, measured in terms of energy content degraded into heat, increases always with the passage of time.

From this, and taking into account that the present entropy of our part of the universe is far from maximum, it is possible to infer an origin in the past and an end in the future. Kelvin, in his paper "On a Universal Tendency in Nature to the Dissipation of Mechanical Energy"[18] summarized the consequences of Carnot's law (published in 1824 and almost forgotten by midnineteenth century) as follows: 1) There is a universal tendency in the material world to dissipate mechanical energy; 2) The restoration of any amount of mechanical energy is impossible without producing an even greater amount of energy dissipation; and 3) The Earth would therefore have a limited future period during which it would be fit for human habitation. Helmholtz spoke,[19] in this connection, of the threat of a "day of judgment", happily obscured in the distant future. Later objections to the applicability of the entropy principle to the universe rested on the consideration that it, i.e. the universe, might be infinite. These objections loose force in the light of information provided by twentieth century astrophysical cosmology, which points to a finite observable universe.

A far reaching consequence of the entropy principle, as applicable to the universe, is that, even if the present mass density were much larger than what is commonly accepted, and it were sufficient to close down the present phase of expansion (giving rise to an oscillating universe) this possibility would not prevent the universe from wasting same energy (as degraded heat or photons) in each oscillation. Then the oscillations would be progressively damped, and no return to the previous state would be possible.[20] A somewhat disturbing feature of the standard Big Bang model as usually presented is that the specific entropy of the universe is assumed to be constant (and very high, of the order of 10^{10}, as given by the very large number of photons per baryon) from very early times to the present.

It would be more reasonable, especially in the early high temperature phases of the universe, before "decoupling", to expect either a steady entropy increase, or abrupt increases at specific times along the expansion.

Another interesting implication of contemporary scientific astrophysics is connected with the concept of the "anthropic principle, a term coined by some cosmologists to take note of the striking fact that everything looks as if the physical laws that govern the evolution of the universe had been directed to the goal of making possible human existence on Earth. Of course, an immanentist or a transcendent meaning can be attached to this principle, depending on the antimetaphysical or metaphysical preferences of the man who is pondering upon it. Hugh of Saint Victor (1096–1141), in the christian middle ages, said: "The world was created for man's body, man's body for his spirit, and man's spirit for God: the spirit that it might be brought into subjection unto God, the body unto the spirit, and the world unto the body". This does sound very much as an early, metaphysical version, of the "anthropic principle". In this connection the words of Prof. R. Jastrow,[21] after pointing out that changes by a few percent in the strength of the fundamental forces in either direction would have prevented the coming into existence of the basic ingredients of life, may be brought to consideration: "Thus, according to the physicist and the astronomer, it appears that the Universe was constructed within very narrow limits, in such a way that man could dwell in it". This result is called the "anthropic principle". It is the most theistic result ever to come out of science, in my view." And he goes on: "Some scientist suggest, in an effort to avoid the theistic or theological implications in their findings, that there must be an infinite number of universes, representing all possible combinations of basic forces and conditions... But I find this to be a typical theorist's solution. In any case, it is an untestable proposition because all there other universes are forever beyond the range of our observations. What is forever unobservable and unverifiable seems to me to be scientifically uninteresting".

It may seem there is some similarity between what happens with the anthropic principle, somehow inscribed into the broad features of the physical universe, and what happens when one looks to one of these pictures portraying a blurred mountainous landscape within the contours of which the profile of a human face can be detected, under close attention. The human face can be easily overlooked, but once you have seen it, it remains unmistakably there.

6.4. The Physical Universe and its Creator

The term "scientific cosmology" refers to the body of human scientific knowledge about the totality of observable matter and energy in interaction, and about the underlying physical laws which appear to rule it. On the other hand, the term "transcendence" makes reference, in a metaphysical sense, to a transcendent God, Creator of both the physical and spiritual world, who sustains into existence all matter and all energy, but who is Subsistent Spirit himself. The presence of the Creator extends itself to all existing realities, physical and spiritual, but it wholly transcends them without becoming alien to of them. The three great monotheist religions, Christianism, Judaism and Islam, affirm clearly this transcendent God, omnipotent and merciful at the same time, and affirm also that He is cognoscible not only by Revelation but also by the right use of reason, from the perfections observable in what He has created. The great religions of the East (Brahmanism, Buddhism, Confucianism), as well as those of other civilization now extinct, do not set forth in such clear terms the existence of a transcendent Creator, even if they contain valuable elements capable of rising men above the purely instinctive and animal level, and pushing him to unselfish wisdom and to love for his fellow men.

Since physical cosmology is about material reality, and transcendence is about a spiritual God, there should seem to be no effective connection between both, at first sight. But this is not the case, because, as shown elocuently[22] by Prof. S. L. Jaki in several books on this subject, there is a subtle connection between scientific cosmology, epistemologic realism and healthy natural theology. In fact, the incipient scientific and technologic developments in many great, now extinct or dormant civilizations, have been followed by premature stillbirths. Science did not achieve in any of them sufficient maturity as to be able to get to the stage of selfsustined progress, characteristic of our christian western civilization from the closing centuries of the Middle Age and the beginning of the Modern onwards. It is a fact of history that science, and scientific cosmology in particular, has developed to maturity only in the cultural ambience of the christian civilization, even if also is true that it became soon emancipated of its original Christian roots.

The statement by a prominent contemporary physicist to the effect that "the truly extrange thing is that there exists something" is a good starting point to make the great leap forward from things created to Creator. The fascination of man for the splendor of the stars in the night sky, turning quietly and unrelentlessly around him, is as old as man himself. Not surprisingly, the beauty, stability and grandeur of the starry realm is often occasion for men to reflect on a higher majesty and stability, that of the Creator. As the medieval christian philosopher would put it into a nutshell: "there are beings which are "contingent" (i.e. they can exists or not); but no contingent being has in itself the reason for its existence; then there must be a Necessary Being who has in itself the reason for the existence of contingent beings. If at any time there was nothing, nothing could be now in existence." The contingency of things in the physical universe is well attested by modern cosmology, which sets forth definite life-spans (even if they are counted in terms of thousand millions of years) for stars, and galaxies, and even for the universe itself.

The pioneers of science (or most of them, at least) have been specially sensitive to the order prevailing in the cosmos. The record is very impressive in this respect.[23] Copernicus, Kepler, Galileus, Newton, Euler, Maupertuis, Joule, Mayer, Helmholtz, Oersted, Ampere, Henry, Faraday and Maxwell, among other lesser figures, have left for posterity eloquent testimonies attributing this order to the Creator. As a sample we may remember the words of Sir Isaac Newton in his "Principia"[24]: "This most beautiful system of the sun, planets and comets could only proceed from the counsel and dominion of an intelligent and powerful Being." Similar words can be said today to do justice to the most beautiful system of the entire universe formed by myriads of radially expanding galaxies made up of millions of glowing stars.

It is well known that for Immanuel Kant universe soul and God are bastard products of the metaphysical cravings of the human mind. Contemporary science, having validated through General Relativity the concept of one universe as the totality of coherently interacting matter and radiation, shows how scientifically unfounded was this categorical statement of the Koenisberg philosopher.

6.5. God and the Scientists

God's existence is a fact much more basic than the fact of science itself, because God is the "Unum Neccesarium", while science is "contingent". This fact is open, by its own nature, to the common sense of every human being, be this a scientist or not. The ground to conclude God's existence, on the other hand, is the wonder and astonishment produced in our intellect by the contemplation of his handiwork (universe, life, man). And in this sense, contrary to what some distinguished science popularizes would like to make us believe, modern science can contribute considerably to widen and deepen our wonderment before the symphonic order of created realities.

It is relatively easy to supplant or to forget God. It is not so easy to deny his existence with a minimum of rational coherence. History is an eloquent witness of the fact that the great majority of men and women of all times have recognized, in their own way, the reality of the divine.

Often, it is true; this natural recognition of the divine appears shaded by fatalistic conceptions of world and life, what is understandable if reason is not supplemented by Revelation. But it is an indisputable fact, which should not be allowed to go under the rug without further comment.

There are two possible alternatives to the recognition of the existence of a Creator. One is "solipsism", the other "pantheism". The first helds that outside my thinking nothing exists. To a solipsist that, on certain occasion, was trying to convince G.K. Chesterton that he was right in his philosophical position, the great humorist replied: "Cherish it!". On the other hand pantheism helds, in short, that everything is in everything, or everything becomes everything. It is true that there are analogies between the different realities with which the human intellect comes into contact; but analogy presupposes diversity, differentiation. On one hand the independent objective reality of hard physical facts (no matter what some physicist would like to make of Heisenberg's uncertainty principle) and, on the other hand, one's own primary experience of moving at will one's own finger, contradict the basic tenet of pantheism. This does not mean, of course, that some great minds have not been attracted to pantheism to solipsism (sometimes both extremes touch each other). But eve then, when they do creative science, they act as true realists.

Scientists in their relation to their Creator are usually not very different from ordinary men surrounding them at the same place and epoch. Copernicus, Kepler and Galileus were devout christians. Newton and Leibnitz entertained no doubt about God as a powerful and provident Creator. In the nineteenth century some great scientists, like Ampere and Maxwell, were practicing believers in the God of creation as well as revelation. Others leaned to materialism, or at least skepticism. Planck and Einstein, the two great pioneers of physics in the twentieth century, were strikingly sensitive[25] to the wonderful ineligibility of the physical world, in spite of the materialistic environment of european scientific circles in which they grew up at the turn of the previous century. And, if the physical universe is well made, and the human intellect well suited to investigate it, it is not so difficult to conclude that this is so because both are the handiwork of a wise and provident Creator.

As we approach the end of the present century, the success of physical cosmology in giving a coherent picture of the universe stands out as a great achievement, even greater than the many technological achievements of which our age can rightly be proud of. In words of Prof. Murray Gell-Mann: "Most remarkable will be that a handful of beings on a small planet circling an insignificant star will have traced their origin back to the very beginning — a small speck of the universe comprehending the whole".

To recognize this is certainly something very important. Still better would be to recognize that an indispensable pre-requisite for this "small speck" to comprehend the whole is the active presence of a transcendent God, creator of both the speck's intellect and the vast surrounding universe.

Bibliography

1. R. Jastrow, "God and the Astronomers", p. 11 (New York: W.W. Norton and Co., 1978).
2. H. Bondi, "Cosmology" (Cambridge: Cambridge University Press, 1952).
3. S.L. Jaki, "Science and Creation" (Lanham, New York: University Press of America, 1990).
4. R. Jastrow, "God and the Astronomers", p. 116 (New York: W.W. Norton and Co., 1978).
5. H.B. Boück, G.V. Coyne and M.S. Longair, eds., "Astrophysical Cosmology" (Vatican City: Pontificia Academia Scientarum, 1982).
6. Ibid., J. Audouze entitled "Primordial nucleosynthesis and its consequences", p. 385.
7. J. Last, "Neutron News", p. 27 (New York: Gordon and Breach, 1991).
8. R.A. Alpher and R. Herman, "Physics Today", 41 (Part 1), 24 (1988).
9. A.A. Penzias and R. Wilson, Astrophys. J. 142, 419 (1965).
10. Various references to the isotropy of the background cosmic radiation in "Physics Today" (April 1991).
11. S. Weinberger, "The First Three Minutes", p. Ill (New York: Bantam Books, 1980).
12. M. Schlik, Nov. 28, 1930. Quoted in the translation of G. Holton, "Einstein, Mach and the Search for Reality", p. 243, in "Thematic Origins of Scientific Thought: Kepler to Einstein" (Cambridge: Harvard University Press, 1973).
13. S.L. Jacki, "The Relevance of Physics", pp. 455–56 (Chicago: The University of Chicago Press, 1973).
14. R.P. Feynman, "The Relation of Science and Religion", in "Engineering and Science" 20, p. 20 (June, 1956).
15. W. Bragg, "The World of Sound", pp. 195–96 (London: G. Bell, Sons, 1920).

16. "The Principle of Relativity Collection of Original Papers on the Special and General Theory of Relativity", by H.A. Lorentz, A. Einstein, H. Minkowski and H. Weyl, with notes by A. Sommerfeld, translated by W. Perret and G.B. Jeffrey, p. 187 (New York: Dover, 1952).
17. Section 3.1 in Chap. 3, and note 1 in this Chapter.
18. W. Thomson, "An Account of Carnot's Theory of the Motive Power of Heat" (1849), in "Mathematical and Physical Papers I", p. 119 (Cambridge: 1882).
19. H.L.F. von Helmholtz, "On the Interaction of Natural Forces", in "Popular Lectures on Scientific Subjects" translated by E. Atkinson (New York: 1873).
20. R.C. Tolman, "Relativity, Thermodynamics and Cosmology" (Oxford, Claredon Press, 1934).
21. "The Intellectuals Speak Out about God", R.A. Varghese Ed., p. 22 (Chicago: Reguery Gateway, 1984).
22. S.L. Jaki, "Science and Creation" (Lanham, New York: University Press of America, 1990) and "The Road of Science and the Ways to God" (Chicago: The University of Chicago Press, 1978).
23. S.L. Jaki, "The Relevance of Physics", Chap. 10 (Chicago: The University of Chicago Press, 1966).
24. "Sir Isaac Newton's Mathematical Principles of Natural Philosophy and His System of the World", translated by A. Motte; revised and edited by F. Cajori, p. 544 (Berkeley: University of California Press, 1934).
25. S.L. Jaki, "The Road of Science and the Ways to God", Chaps. 11 and 12 (Chicago: The University of Chicago Press, 1978).
26. "National Geographic", 167, p. 662 (May 1985).

The Cosmic Background Radiation
(El Escorial 1993)

Chapter 7

The COBE Project, by John C. Mather

To begin with I would like to point out that we can look back in time by looking at things that are very far away. We simply use the fact that light comes to us at a definite speed and so, when we look at the sun, we see the sun as it was 500 seconds before we look at it; when we look at something in the middle of the galaxy we see it as it was 25,000 years ago and when we look at something extremely far away we can see it as it was 15 billion (10^9) years ago.

Hand		1m	0,000.000.003 sec
Earth		7.000 Km	0,02 sec
Sun		150.000.000 Km	500 sec
Stars			4 years
Galaxy			25.000 years
Big Bang			15.000.000.000 years

Fig. 7.1. Hand, earth, sun, stars, galaxy and Big Bang.

We have various ways to measure the size of the universe (Fig. 7.1) and to learn about the Big Bang. Very briefly: We measure distances by classical geometrical methods drawing triangles (Fig. 7.2) and comparing the brightnesses of standard candles. Much of the effort of astronomy for many generations now has been concerned with determining what makes a candle a standard candle, how we can recognize distant objects as being the same so that we can determine relative distances by comparing the brightnesses.

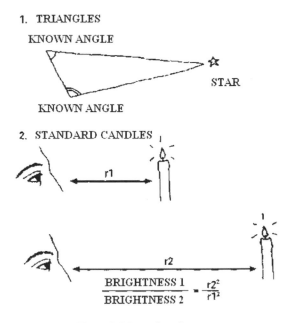

Fig. 7.2. Measuring distances.

We measure distances in this way and we measure velocities, in most cases, in only one way, what we call the Doppler shift (Fig. 7.3).

Using this effect, you can determine how fast distant objects seem to be going away from us. And this was the basis of Hubble's discovery in 1929 that distant galaxies recede from us with a speed proportional to the

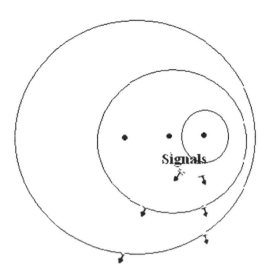

Fig. 7.3. Measuring velocity: A wave source in motion produces waves whose lengths depend on the direction and the velocity of the source.

distance. At first, the numbers that Hubble used for the distances were incorrect, very incorrect, so it was thought that the universe was only 2 billions years old. Nowadays people think that the universe is about 15 billion years old, but it is still a subject of great controversy. Hubble was only able to study the smaller part at the graph in the lower left corner of the figure (Fig. 7.4). Now we have measurements that go to much wide distances and the basic picture is very simple: there is a perfect straight line here and the deviations from perfection are very small. It's immediately possible to conclude from this that the universe has no observable center that we can discover.

We have here (Fig. 7.5) three astronomers, all look out at the sky and see Hubble's law. We imagine we are not in a special place in the universe and we can actually conclude from the fact that Hubble law is a perfect proportionality that every astronomer would deduce the same speed of expansion for the universe, and each of these astronomers would conclude that he was at the centre of the universe. Clearly this in incorrect, we can not be all in the center of the universe. We can,

HUBBLE

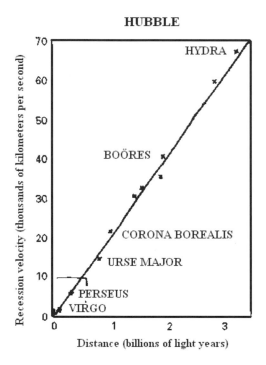

Fig. 7.4. A representative velocity-distance curve showing the expansion of the universe. The lower left corner is the region surveyed by Hubble up to 1929.

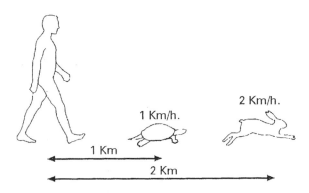

Fig. 7.5. Hubble's law no necessary center!

however, conclude that the universe is so large that we cannot find the center, if there is one. Now, we would like to know what the universe is going to do next. So we must think about the law of gravity. Gravity attacks distant galaxies to each other (Fig. 7.6).

This is a very difficult thing to think about if you imagine the universe is infinite. Isaac Newton thought about this and realized that it was impossible to completely understand it with his theory of gravity because too many of the forces become infinite. So, we need Albert Einstein's theory of general relativity before we can't really understand what distant galaxies pulling each others do.

As a result we can draw a curve which shows three possible histories that the universe could have (Fig. 7.7). On the horizontal axis we have the tinte since the apparent explosion and in the vertical axis we have the apparent size of the universe. We can imagine three possibilities: One is that the expansion continued for a while, then turned around and failed back in a great cosmic crunch; second is that the universe will continue to expand towards a zero final velocity, and the third is that it will continue to expand to a final velocity that will not be zero but a definite number and it will continue to expand quite rapidly. We are presently unable to determine which of these conditions is going to be the case. We already talked about the lambda constant and the amount of mass in the universe. Measurements of those numbers are necessary to deduce which of these is our probably future.

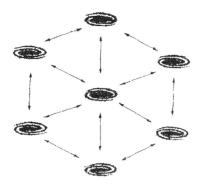

Fig. 7.6. Galaxies attract each other.

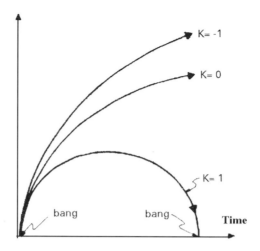

Fig. 7.7. Friedmann universes all begin with big bangs.

One of the great mysteries of astronomy at this time is about how much material in the universe (because of the gravity) will be necessary to stop the expansion. We have three different ways now in which we can explain this unknown material that is not visible to us (Fig. 7.8).

The first method is that when we watch how galaxies rotate, we don't observe that they rotate as the stars would rotate around a mass point in the center of the galaxy; in this case the velocities would come down as the distances of the stars from the center of the galaxy increased. But what we actually see is that the rotation velocity achieves a definite value and stags very high over very relative great distances of the center of the galaxy. And when you add up all the mass that must be included to explain the rotation curve, we conclude that the mass must be infinite. This is quite unlikely (clearly we cannot extend the integration to this level) but it does mean that there is very strong evidence even from objects of the size of normal galaxies, that there is a great deal of missing material.

The second way that has been demonstrated recently is that clusters of galaxies which contain hot gas in between the galaxies themselves emit X-rays and these X-rays are characteristic of a particular temperature of

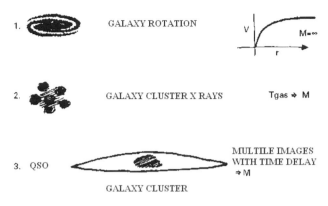

1. GALAXY ROTATION

2. GALAXY CLUSTER X RAYS   Tgas ⇒ M

3. QSO   MULTILE IMAGES WITH TIME DELAY ⇒ M

GALAXY CLUSTER

Fig. 7.8. Dark matter.

the gas in that space. From the temperature of that gas and our understanding how the gas is heated we can immediatcly deduce how much material is in the cluster of galaxies. We see how the galaxies move in the gravitational field, we see the temperature of the gas and it all comes out with the same estimate, that there is a great deal more additional material than the luminous material in the stars. In some clusters there is ten tunes as much invisible material as there is in stars.

The third and extremely convincing method to deduce the masses of galaxy clusters is by studying images of distant quasars. It turns out that the gravitational field of the clusters of galaxies can be sufficiently strong that light can travel in different paths around a galaxy cluster to reach ourselves as we are observing, and we see several separate images of the same quasar. We can demonstrate that it is the same quasar because there are many identical things about this two images, and, from that, we can measure the mass of the cluster of galaxies in the middle. In all three of these ways, we are quite sure there is some kind of dark material that is not observable as ordinary matter.

All of these leads us to a grand picture of the Big Bang and a wonderful and complex story about how it might have happened, and it's quite remarkable that so many people have come to agree that this is the true story, and not everything of this can be proved.

In the beginning we imagine that there was a primordial material which, for unknown reasons, existed. We are unable to establish what was there before that, and this is not even a meaningful question for us. For some reason, this material began an exponential inflactionary period, as has already been described.

Energy is released from some submacroscopic force which drives an expansion of the actual space. Initially, a ball of material which is presumed in equilibrium at a single temperature, expanded into a space that it is so large that individual particles are unable to receive light from another. As the expansion proceeded, what we call scale free quantum fluctuations were produced (Fig. 7.9), and they are going to be later the things that produce galaxies. At this time there were many kinds of particles and antiparticles and most of them disappeared producing photons. Now it is possible that in this phase transition, that is like the freezing of water, or the evaporation of water, there would be domains boundaries. It is possible that the expansion would carry them away so far away that we cannot find then. Around 10^{-15} sec. after that explosion is the time at which we think the laws of physics were established. But there are good reasons to think that there was additional dark matter that we don't understand, so even at this point we really don't know the laws of physic.

As the expansion continues there are several additional events. We all know how about three minutes after the Big Bang nucleosynthesis begins. Other major events are given in the accompanying table (Fig. 7.10). Dr. Smoot already showed you the next picture (Fig. 7.11)

PRIMORDIAL SOUP – UNIFIED FORCES- QUANTUM GRAVITY

EXPONENTIAL INFLATION- FREEZES IN EXISTING FLUCTUATIONS

TAKES INITIAL REGIONS OUT OF CAUSAL CONTACT

PRODUCES SCALE-FREE FLUCTUATION SPECTRUM

MATTER (ALL TYPES), PHOTONS ON EQUAL FOOTING

INFLATES AWAY TOPOLOGICAL DEFECTS FROM PHASE TRANSITIONS

10^{-15}sec., T=100 GeV, Known physics?

?-DARK MATTER DECOUPLES, BEGINS TO MOVE WHEN CAUSALLY CONNECTED

JOHN MATTER 301-286-8720 MATHER@STAR.GSFC.NASA.GOV

Fig. 7.9. Major events in cosmic expansion.

<u>3 MINUNTES</u> – LIGHT ELEMENT NUCLEOSYNTHESIS

<u>1 YEAR (Z=106.5)</u> – FREEZE PHOTON NUMBER

-ALLOW CHEMICAL (BUT CMBR IS DOMINANT)

$\eta=1/(e^{x+\mu}-1)$; μ Potential

<u>1000 YEARS</u>-EQUAL DENSITY OF MASS AND PHOTONS

ALLOW MIX OF TEMPERATURES OR HEATING BY ELECTRONS

$\eta=<1/(e^x-1)>$; $x=h\nu/kT$

Y=INTEGRAL OF ELECTRON PRESSURE

<u>300,000 YEARS</u>- DECOUPLING-RADIATION OBSERVED AS IT WAS THEN

NOT CASUALLY CONNECTED FOR >2 DEGREE SCALE

ANISOTROPY FROM GRAVITATIONAL POTENTIAL, DOPPELR SHIFT

$\sim 10^9$ YEARS-GALAXY FORMATION + CLUSTERING

Fig. 7.10. Other major events.

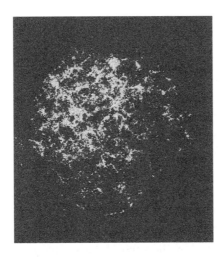

Fig. 7.11. Illustration generated by computer of the location of a million galaxies.

to indicate that the universe is not the same in every direction and this is one of the great motivations that we had for proposing to build the COBE. We knew that this pattern existed. This is a computer-generated drawing of the location of a million galaxies. They are not homogeneously distributed throughout the sky. This is impossible without a primordial structure.

So we designed the COBE satellite with three mayor objectives (DMR/FIRAS/DIRBE) (Fig. 7.12). The satellite was launched on 11/18/1989 by a Delta Rocket after fifteen years of study, design and preparation, mostly at the Goddard Space Flight Center.

DEFINITIVE MEASUREMENTS OF TWO COSMOLOGICAL FOSSILS:

1. COSMIC 3K MICROWAVE BACKGROUND RADIATION

   – LARGE ANGULAR SCALE ANISOTROPY

   – SPECTRUM (100 μm- 1cm)

2. COSMIC INFRARED BACKGROUND RADIATION (1-300 μm)

LAUNCH 11/18/1989

DELTA ROCKET

Fig. 7.12. COBE objectives.

George showed you this list (Fig. 7.13). Many people were here busy all these years to make it happen. I will not go into the list of exactly who did what.

Let me point out the drawing (Fig. 7.14) of the spacecraft and illustrate the important features of it. There is a liquid He cryostat that at launch carries 600 litter of liquid He. Outside there are three boxes which carry the microwave receivers which obtained the data which George described to you. There is a protective shield which keeps the sun and the Earth from shining on the instruments. The spacecraft with all of the electronic boxes, power controllers and transmitters, are all in the bottom half, and all is powered by the solar electric panels.

The Earth is always maintained bellow the satellite and the sun always shines in from a side. We put the satellite in the orbit indicated in the figure (Fig. 7.15) which shows the orbit in the day the satellite was launched, from California towards the South in the direction indicated. The orbit plane is approximately perpendicular to the line from the Sun but not exactly. The orbit would actually precesses at approximately one rotation per year, so that this general picture in which the orbit is perpendicular to the line from the Sun is preserved all year long.

BENNETT, C. L.	NASA-GSFC	DMR DEPUTY P.I.
BOGGESS, N. W.		
CHENG, E. S.	NASA-GSFC	
DWEK, E.	NASA-GSFC	
GULKIS, S.	NASA-JPL	
HAUSER, M. G.	NASA-GSFC	DIRBE P.I.
JANSSEN, M.	NASA-JPL	
KELSALL, T.	NASA-GSFC	DIRBE DEPUTY P.I.
LUBINM, P. M.	U.C.S.B	
MATHER, J. C.	NASA-GSFC	FIRAS P.I. & COBE Project Scientist
MEYER, S. S.	U. Chicago	
MOSELY, S. H.	NASA-GSFC	
MURDOCK, T. L.	GEN. RES. CORP.	
SHAFER, R. A.	NASA-GSFC	
SIVERBERG, R. F. L.B.L.	DMR P.I.	
VRTILEK, J.	NASA-HQ	PROGRAM SCIENTIST
WEISS, R.	M.I.T.	SWG CHAIRMAIN
WILKINSON, D. T.	PRINCETON	
WRIGHT, E. L.	U.C.L.A	

COBE DESIGNED BUILT, + LAUNCHED, AND DATA ANALYZED BY NASA GODDARD SPACE FLIGHT CENTER

Fig. 7.13. COBE, science working group.

Let me illustrate some of the things we want to study with the COBE. This is a graph of the brightness of the radiation that we could find over the range of radiation from 1 cm to 1 μm (Fig. 7.16). At the long wavelength end is the cosmic background microwave radiation from the Big Bang. But there are many other things that also irradiate at the infrared and microwave radiation, the brightest of which is the interplanetary dust. This is made up of dust grains in between the planets, constantly orbiting around the debris Sun, and spiralling in towards the Sun. They are the debris of asteroids colliding with one another, and from comets. There is also starlight from our galaxy. There are infrared galaxies, already catalogued, and they are predicted to give a brightness

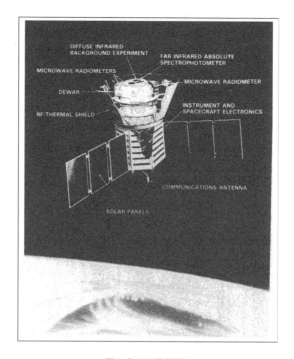

Fig. 7.14. COBE.

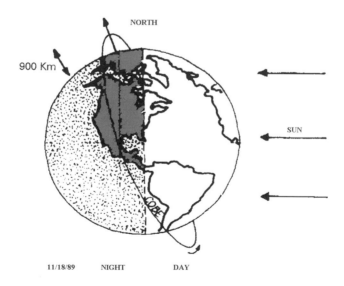

Fig. 7.15. Drawing: North, night, day, sun.

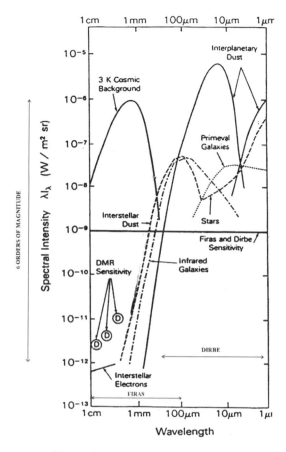

Fig. 7.16. Wavelength vs. spectral intensity.

about that indicated, and there is interstellar dust. About a fourth of all the radiation from the stars in our galaxy is absorbed by interstellar dust grains and re-irradiated at far infrared wavelengths.

Finally, we have here a curve which shows a prediction for the possible brightness of primeval galaxies as they fill up the universe with infrared radiation. It is important to point out that some of the things we want to study, like the 3°K cosmic background radiation (CBR), are very bright, relative to the things that come out from other local sources. The CBR is about 10^6 times brighter than the interstellar dust radiation. That means we can get a very clear view of the primeval cosmos at wavelengths around it. Other main objective, to find the primeval galaxies, is much more difficult, because even the optimistic prediction shown indicates that only at about a few wavelengths ($\approx$ 3 µm) are we likely to find their radiation easily. At other radiations we must first understand the radiation from interstellar and interplanetary dust to be able to see radiation from primeval galaxies if it exists at all.

I will show you the concept for the spectrometer experiment. This instrument was intended to test whether the cosmic background radiation (CBR) has a black-body radiation field as it is predicted to have in a simple Big Bang picture. It is a Michelson interferometer operated as a spectrometer (Fig. 7.17). I will show how this instrument operates and how it is calibrated. The light from the sky comes down to the concentrating antenna and bounces off a mirror, and comes into the interferometer. There is a beam splitter which divides the incoming light equally into two parts going in perpendicular ways, and coming back. When the beams encounter the beam-splitter again the y recombine and

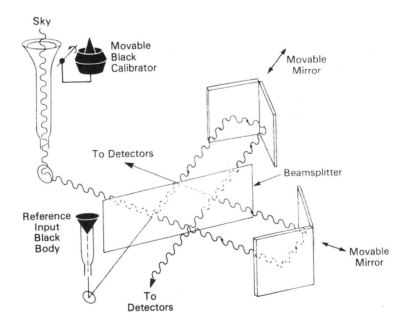

Fig. 7.17. Instrument.

the amount of radiation which goes this way or that way depends on how much different the lengths of the two arms are. There is one position at which all the waves add up coherently and go to a single detector. In other positions the intensity depends on the wavelength.

The instrument is going to be calibrated by a black object, which is a simulator of the universe at the Big Bang, ideally a non-reflective object, whose temperature we can control to match the temperature we now observe for the CBR. To make this instrument more accurate we operate it in a differential way. We have a second black-body as a second input and we use this also and we see this also to match the temperature of the CBR. If the radiation from the sky and the radiation from this reference object have the same brightness, when the mirrors over there move, there is no change in intensity going to the detectors, so this makes the interferometer a differential instrument. The detectors that use are very sensitive thermometers calls bolometers.

Here is an example of data taken with this instrument as the mirrors are moved (Fig. 7.18). The horizontal axis shows the distance the mirrors have moved in arbitrary units (we count the marks on a scale). The vertical axis is the intensity. In the first graph (upper graph) we have shown our first measurement of the actual sky. You see that the curve is almost level all the way across. Then, we said, let us calibrate this instrument by changing the temperature of the internal reference body. We heated up a bit and we got the next quite different curve. Then we said let us also put the external calibrator (the one that simulates the big-bang) in the antenna beam, and we got next curve. So we were able to adjust the temperature and conclude, that just by looking at these charts the CBR has a spectrum that is very similar to that from a black-body at a temperature of 2.7 K.

To make this more precise we have engaged in several years of calculations to understand the calibration of the instrument and now I am going to show you some of our results (Fig. 7.19).

This is a map of the temperature of the sky and it looks very much like the one Dr. Smoot showed you on Monday. But this is made with an absolutely calibrated instrument, and it is made separately for each direction. So, in order to make this map we have taken the original

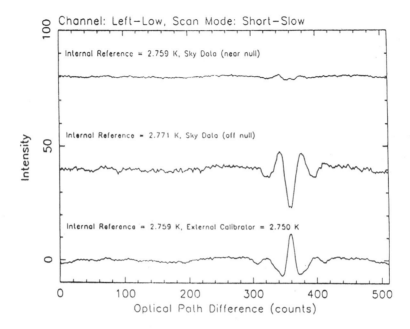

Fig. 7.18. FIRAS interferograms.

Fig. 7.19. Map: COBE FIRAS CMBR temperature.

spectrum from the instrument an we have substracted a model of the emission from dust in our galaxy. Although the middle of the galaxy is right here you see very little trace of it. This means that we understand also interstellar dust very well. Just as Dr. Smoot told you this part of the sky is a little brighter (3.3 mK brighter), this other part 3.3 mK fainter. This is due to the motion of the Earth relative to the microwave CBR.

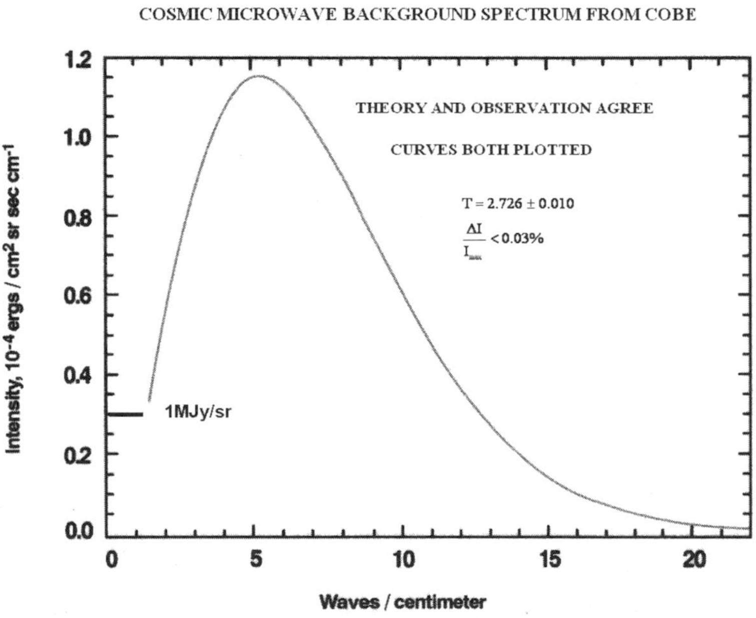

COSMIC MICROWAVE BACKGROUND SPECTRUM FROM COBE

Fig. 7.20. Cosmic microwave background spectrum from COBE.

Here is the result (Fig. 7.20), in numerical terms, of the spectrum. On this chart we have both theoretical and experimental data plotted, and error bars in the data are too small to show. The temperature of this black-body radiation is 2.726 ± 0.010 K. The maximum difference between the perfect curve and the observed data is only 0.03% of the brightness. This is as close to perfection as we are likely to get for a long time in a measurement of a black-body.

To show you a little more precisely what the differences are, here is a chart (Fig. 7.21) where we have expanded the vertical scale and we have substracted the theoretical model from the observations. You see a straight line across the middle of the chart. It shows that the microwave CBR has a spectrum that is indistinguishable from a black-body at all these wavelengths down to this very, very small error bar (at the left hand side). The immediate conclusion is that the Big Bang was very simple.

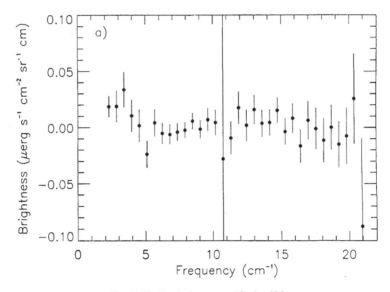

Fig. 7.21. Deviation magnified ×600.

Nothing much happened to changed it. To be more quantitative about that, let me shown a sketch of the things that might have happened to the black-body radiation (Fig. 7.22). Suppose that there was a process in the early times which liberated energy. As an example supposes that there were black were holes which were consuming matter and generating a lot of energy that could be added to the CBR. This would work by producing hot electrons, and the electrons can bounce off the photons and change their wavelength. How much change occurs depends on how fast electrons are moving. So it is actually possible to change the energy in the CBR field without changing the number of photons. If that happens the CBR cannot have a black-body field. Another process that can occur is that an electron that passes very close to a proton can actually create a photon by scattering in the electromagnetic field of the proton. This happens

COMPTON SCATTERING

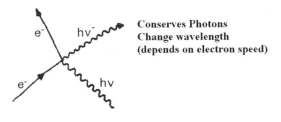

Conserves Photons
Change wavelength
(depends on electron speed)

FREE-FREE EMISION

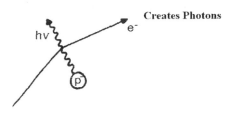

Creates Photons

Fig 7.22. Compton scattering and free-free emission.

primarily at very long wavelengths. So, as to how we do the calculations about this, we can conclude that there are several possible mathematical forms, which we don't have to remember, for kinds of distorsions of the spectrum.

I will just mention how we interpret them. The first one is that if the temperature of the radiation has been changed then we get a particular kind of spectrum form. This first is the derivative of Planck's function with respect to temperature and ΔT is the change in temperature (Fig. 7.23).

We actually don't know what the temperature of the radiation was supposed to be, so this one does not have cosmological meaning for us. The second possibility is that early energy release would have created what we call a Bose-Einstein distribution with a chemical potential, μ, and we want to have a limit on the value of μ in order to limit the energy release in the early universe. The third possibility is the one I just sketched to you, where the hot electrons change because they mix their temperature and this has a completely different form for the shape of this distorted spectrum. And finally (fourth) we have the possibility that there is dust in our own galaxy and then this has yet a different spectrum. In the end we have a chart that shows (Fig. 7.24) how much energy could have been released by various forms of exotic processes as a function of the timc of expansion. On the vertical axis I have the ratio between an energy released and the total amount of energy in the CBR, and on the horizontal axis I have the redshift which is also the ratio of the size of the universe now to the size of the universe at the time this event might have occurred. So you see that over a very wide range of redshift $\Delta U/U$ is very small and then back in time the limit is not quite as tight, and then this curve rising up here $(1 + z > 10^6)$ tells us that we cannot see back beyond about the first year. So this basically summarizes the fact that the Big Bang really was very simple.

1. TEMPERATURE SHIFT

$$\frac{\partial B_v}{\partial t} \Delta t$$

2. CHEMICAL POTENTIAL

$$\frac{2hv^3}{c^2}\left(\frac{1}{e^{x+\mu}} - \frac{1}{e^x - 1}\right) = \frac{T}{x}\mu\frac{\partial B_v}{\partial T}$$

3. COMPTON, TEMPERATURE MIX

$$y \cdot T_o \frac{\partial B_v}{\partial T}\left(\frac{x(1+e^x)}{1-e^x} - 4\right)$$

4. DOST $T\left(\dfrac{v}{v_o}\right)^2 B_v(T)$; MIX OF TEMPERATURE OR

$$T\left(\frac{v}{v_o}\right)^\alpha B_v(T) \text{ OR ELABORATE CALCULATION}$$

Fig. 7.23. Spectrum distortion.

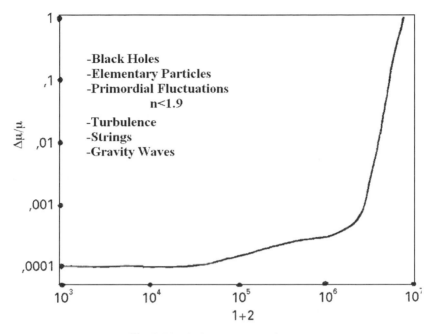

Fig. 7.24. Limits on energy release.

What can we say knowing that the Big Bang was so simple? There are a number of alternate theories that one can examine. One that has been already discussed a lot was the Steady State Theory, with a continuing expansion, and no single explosion at the beginning (Fig. 7.25). The problem with this idea is that radiation had to be generated continuouslly. The whole idea of the Steady State is that it is continuous, and matter is being generated to replace the matter which has expended away. The trouble with this model is that it cannot easily produce a blackbody spectrum. The radiation which we receive has to come from a wide range of distances, a wide range of redshifts, and it is almost impossible to make a perfect black-body spectrum from adding up things that are not blackbodies. A similar problem occurs if you imagine that the Big Bang was cold. This is one of the clever alternatives, but the trouble is very much the same. The radiation is not generated all at once in an equilibrium condition but it is instead generated by adding up many small pieces and it is hard to make a perfect black-body spectrum out of that.

STEADY STATE = CONTINUING EXPASSION

THERE WAS NO BIG BANG

RADIATION IS GENERATED CONTINUOSLY

MATTER IS GENERATED TO REPLACE THE DILUTION

PROBLEM: BLACKBODY SPECTRUM TOO PERFECT

COLD BIG BANG = SAME PROBLEM

Fig. 7.25. Theories tested.

So this leads us to the question of what could that intergalactic material be, if there is some (Fig. 7.26).

NOT HOT – WOULD DISTORT THE SPECTRUM

NOT DUSTY – WOULD PBSCURE VIEW

COULD BE COOL IDNIZED STUFF

COULD BE MASSIVE NEUTRINOS FEW e.V.

Fig. 7.26. What's the inter-galactic stuff?

We know that it is intergalactic material, then it cannot be ordinary matter at a great temperature, because otherwise we would see it. Even now X-rays would be produced from hot material, and the CBR would not be right. It cannot have very much dust in it, because the dust would obscure our view and we know that we can see quite far away back to very early times. It is still possible for it to be a cool ionized material, relatively cool, so that we would not be able to see it. And it is also possible that the dark matter is made up of a massive kind of neutrinos (of a few electron-volts of mass per neutrino). If it was much more than a few electron-volts there would be too much mass, and it would have stopped the expansion of the universe.

It is also possible from the perfection of this blackbody spectrum to conclude that early galaxies cannot be very dusty (Fig. 7.27). And that seems unlikely because dust is very easily generated by stars, as in supernovae, and if early galaxies had dust in them, then a great deal of luminosity would be converted into the very far infrared and we would have seen that in our data. So we can conclude that less than 0.05% of hydrogen has been burned up so far. The universe has a long time to go before we ran out fuel. Our Sun is probably only going to live another ten billion (10) years. In this way the universe as a whole would have plenty of time so that there would be many stars that would last longer than that.

DISTANT, EARLY GALAXIES LOOK LIKE THOSE NOW, BUT BLUER

DUST IS FORMED BY SUPERNOVAE VERY QUICKLY

LUMINOSITY CONVERTED TO IR BY COMBINED DUST AND RED-SHIFT

Fig. 7.27. Early galaxies?

Another theory we can test is whether galaxies would form as a result of some supermassive object that originated in some previous generation of events (Fig. 7.28). It is possible to imagine that something (although we have never seen any example) which might have been much more massive than a galaxy, would have formed something like a giant star, which would have then exploded. If it would have exploded it could have created a giant cavity, and the material would be flaying away from this cavity, and possibly piling up at the outside edges where it would generate galaxies on the surface of the bubble. As Dr. Snoot explained to you we think it is possible that the structure of the universe as a whole is something like a sponge with giant holes and with all of the galaxies concentrated on sheets of material. So, the advantage of this idea is that it would actually generate a distribution of galaxies similar to that we see. The disadvantage would be that the energy liberated by these processes would destroy the CBR spectrum, and we would have known about it.

Advantage: Resembles real spongy distribution of galaxies
Disadvantage: Explosion would distort CMBR spectrum

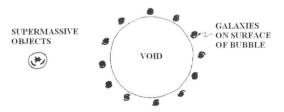

Fig. 7.28. Explosive formation of galaxies.

Let me move to another topic (Fig. 7.29). This one is a measurement we made with our instrument to test whether the CBR has the exact

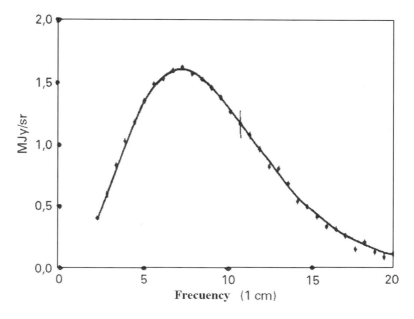

Fig. 7.29. Dipole fit Amp = 3.34 mK.

Doppler interpretation for the dipole. It was earlier said that we think we are moving through the universe at a speed of about 360 Km/s, and that causes the microwave CBR to have a different brightness m each different direction. This is a measurement that shows that it has the right amount of this at every different wavelengths observed in this instrument. Then you see the very small error bars. This is an excellent fit to the data. So it is really proper to consider that this is a Doppler shift effect.

The instrument also turned out to be an excellent probe of the interstellar material (Fig. 7.30). We can see a lot of things with this spectrometer besides the microwave CBR. You see that the sky is full of atoms, ions and molecules. There are carbon monoxide molecules in space in the plane of our own galaxy, and these molecules are seen to

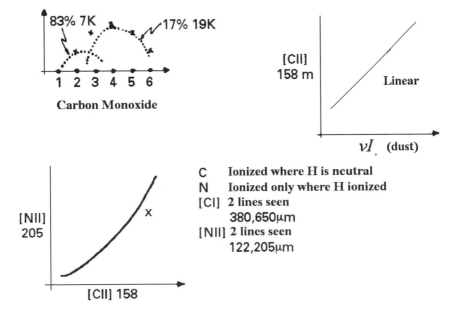

Fig. 7.30. Atoms, ions, molecules.

radiate at several different wavelengths, so we can study the properties of the interstellar gas clouds in that way. We also see there is a carbon ion [CII] that emits at 158 μm, and that the amount of this radiation is about the same, or rather proportional, to the radiation from the dust grains that we see in the same space. So we can learn about the chemistry of interstellar material from this instrument.

One of the most very difficult questions for the astronomers is whether there is cold dust out there (Fig. 7.31). One possibility for the dark matter, that people are talking about, is that there are little small objects, larger than ordinary dust, but smaller than satellites or planets. We have the observational data now to test this theory. We have been able to show that it is possible that there is a significant amount of interstellar dust that is much colder than ordinary interstellar gas.

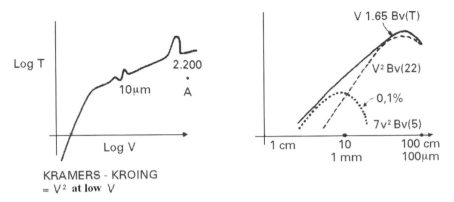

Fig. 7.31. Cold dust.

Most interstellar gas is heated by starlight to a temperature of about 22 K (Fig. 7.32). But also we saw that it is possible to see some dust that is only at 5 K. This is important because it limits our ability to see all the way back to the beginning of the universe. The cold dust would look a lot like the CBR.

For completeness I will repeat a few of the charts that George Smoot showed you, to talk about the anisotropy of the CBR (Fig. 7.33). The measurements that the microwave radiometers on the COBE obtained give us the large angular scale structure of the primordial universe (Fig. 7.34 and 7.35).

We think that it is primordial because different parts of the sky that we observe have not had time to communicate with one another since the beginning of the expansion. So we see (Fig. 7.36) this CBR essentially as it was a microsecond or less after the great explosion.

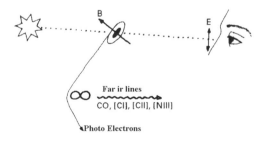

Fig. 7.32. Interstellar dust.

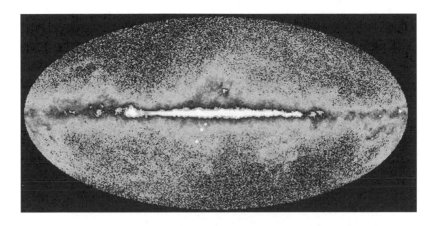

Fig. 7.33 COBE firas.

LARGE ANGULAR SCALE (> FEW DEGREES) IS UNMODIFED PRIMORDIAL STRUCTURE.
INFLATION PREDICTS SCALE FREE POWER LAW FLUCTUATION SPECTRUM.
SCALE INVARIANT = EQUAL GRAVITATIONAL POTENTIALS ON ALL SCALES.
REGIONS NOT CAUSALLY CONNECTED AT TIME OF DECOUPLING
SHOWS GRAVITATIONAL REDSHIFT OF REGION AT DECOUPLING (SACHS-WOLFE EFFECT).
COLD DARK MATTER (IF ANY) CAN GROW STRUCTURES AS SOON AS CAUSALLY ALLOWED.

Fig. 7.34. Anisotropy: Large angular scale.

MEDIUM SCALE (~ 1DEG) BECOMES LARGE SCALE STRUCTURE NOW SPERCLUSTERS, VOIDS.
AFFECTED BY DETAILED PHYSICS AT DECOUPLING.
SMALL SCALE (ARCMINUTES) BECOMES GALAXIES AND CLUSTERS NOW.
ERASED BY DECOUPLING PHYSICS, RE-ESTABLISHED BY GALAXY FORMATION.

SEE INDIVIDUAL CLUSTERS NOW (SUNYAEV-ZELDOVITH EFFECT).

Fig. 7.35. Anisotropy: Medium and small angular scale.

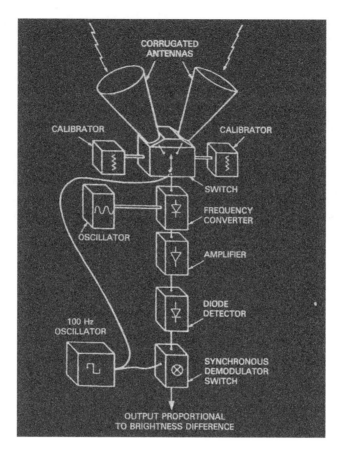

Fig. 7.36. Output proportional to brightness difference.

On smaller angular scales it is possible that the radiation has been modified by various processes since the explosion, because there has been, time for these smaller objects to communicate inside themselves and do other things, to grow into galaxies and clusters and so forth. So there is a great effort currently going on with experiments on airplane balloons and on the ground (Antarctica and everywhere else) to find medium scale anisotropy. Radiotelescopes on the ground can even show a smaller angle scale anisotropy. It is very difficult to do these things, but it is also very, very important.

Yesterday or on Monday there was a question about the Doppler shift and what causes it, and I said that we had a chart showing how the

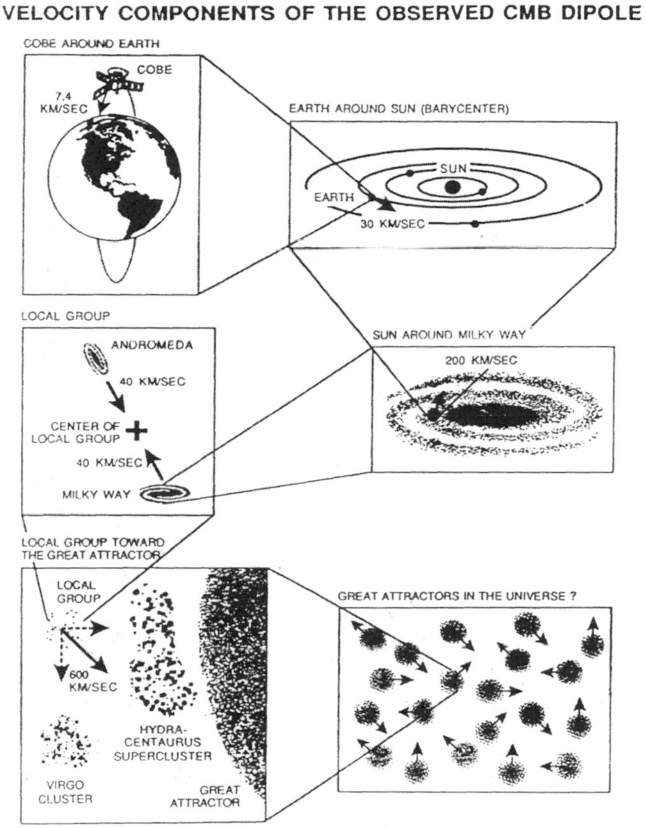

VELOCITY COMPONENTS OF THE OBSERVED CMB DIPOLE

Fig. 7.37. Velocity components of the observed CMB dipole.

velocity that we observe is the vector sum of many different velocities. In the chart (Fig. 7.37), the satellite COBE is orbiting at a speed of 7.4 Km/s. The Earth orbits around the Sun at 30 Km/s. The Sun around the center of the Milky Way galaxy at about 200 Km/s. The Milky Way is falling towards the center of the Local Group at about 40 Km/s. The Local Group of galaxies seems to be attracted towards an object that we call the Great Attractor at a speed about 600 Km/s. And then it is possible that there are many other great attractors in the universe that are pulling on our own local Great Attractor. So we think that the velocity that we measured with COBE, and that it has also been seen before, must be the result of all these processes happening. The most important one for cosmology is this question: whether there are great masses like the great attractor that are able to cause the motion of huge regions of space, as matter in space.

This is the map that this instrument showed (Fig. 7.38). We hope very soon to have ready a new and improved map to show, because we have many more years of data to analyze. This map was made using only one year worth of data from the COBE, and we have now almost completed our fourth year of observation. Although some of the objects that you see on this map are not real because there is noise from the instruments and receivers, you will find when we get our new and better maps that most of the things on the map are really there, and really real.

Now I would like to talk about the third experiment (DIRBE) which is intended to find the energy released from the first generation of galaxies (Fig. 7.39 and 7.40). It is a difficult problem, because as I showed you before, there are many sources of IR radiation besides the infrared CBR. The instrument that we designed to measure this is the only one of the COBE instruments that looks at all like a telescope. It is a very small telescope. The aperture of the primary mirror is only 19 cm. Radiation comes down from sky and is analyzed by the device. It comes from sky and bounces from the primary mirror.

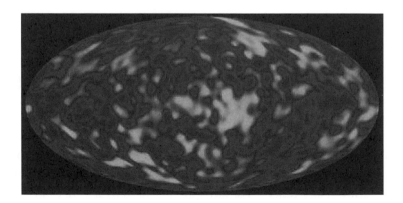

Fig. 7.38. COBE DMR.

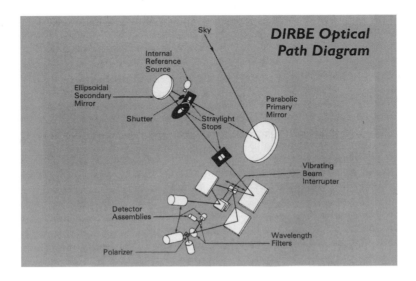

Fig. 7.39. Instrument.

GOAL: DETECT ENERGY RELEASED BY "FIRST GENERATION" GALAXIES

PROBLEM: COMPLICATED AND STRONG FOREGROUND EMISSION

Fig. 7.40. Infrared background.

There is an image from the sky formed here (secondary mirror) and then the image reemits by the secondary mirror and passes down through additional stops and additional focusing mirrors. The light passes through a vibrating beam interrupter, which we call a chopper, and the radiation comes through here, and then it is either bounced off the blades of this chopper into this set of detectors, or transmitted between the blades to this other set of detectors.

The radiation is interrupted 32 times per second, so that we can tell the difference between radiation that came from sky by the correct path, and radiation that may have been originated elsewhere. And this is one of the important parts of our instrument design, that enables us to calibrated the device. Another thing that we can do, which is very important and which is different from other instruments that have been flown, is that we have a shutter here that can block the view of the sky altogether and enables us to look at an internal calibration source. A source of hot radiation that we can control. So this instrument altogether looks at ten different wavelengths over the range from 1 μm to 300 μm, and it observes them all simultaneously. The line of sight of the instrument is inclined relative to the spin axis of the spacecraft so that we rapidly scan out a large portion of the sky in every orbit. The reason for that is that the intensity of the radiation that we want to observe includes the radiation from the interplanetary dust, and that has a different angle from the Sun.

So this is a map (Fig. 7.41) that shows you the observations this instrument made at the short wavelength end of the spectrum. This is a 1, 2 and 3 μm wavelength combined. The blue spots are observed a 1 μm, the red spots are the ones which are bright a 3μm. In the center is the middle of our own galaxy. This is the first time we ever had a very clear view of the center of a galaxy, and you can see these orange spots here that are the interstellar dust clouds, the ones that actually absorb starlight, and prevent us from seeing directly all the way to the center, when we use visible light. The little oval cloud is a collection of mostly redish stars which are called the bulge of the nucleus of our galaxy. So we are actually able to se this even though we live ourselves in the plane of the galaxy and most of the view is obscured by dust.

Fig. 7.41. DIRBE 123 composite.

At a little longer wavelengths we get a different view, and Dr. Smoot has also showed you this already. This chart (Fig. 7.42) now shows you the brightness of the interplanetary dust. This orange pattern is the dust grains in between the planets, and it is at a temperature about the same as you and me, i.e. about 300 K or room temperature, and it is very bright at a wavelength of ≈ 10 μm. You can still see the light from the stars in the plane of our own galaxy. Blue in this picture is a wavelength of 5 μm, yellow or orange is 12 μm, and red, of which we see a few objects, is 25 μm. This shows you the constellation of Orion over here. It is very bright at this wavelength because stars are forming over here in dust clouds and they are actually making a very hot interstellar dust. Another place where this is happening is here. This is the constellation of Cygnus. Many stars are forming in dust clouds there also, producing very much warm dust.

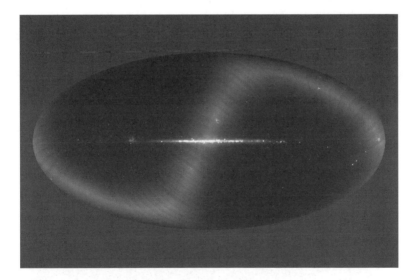

Fig. 7.42. DIRBE 456 composite.

When we look at longer wavelengths the picture changes again (Fig. 7.43). This is a combination of colors of 100 μm, 140 μm and 240 μm. These are such long wavelengths that all we see in this picture is interstellar dust. Here we see the constellation of Orion even brighter.

There is a little speck down here you can barely see which is the Large Magellanic Cloud and all of this you see here is interstellar dust. The same radiation that absorbs at near infrared wavelengths and visible light wavelength is reirradiated at this long wavelength. So far that does not give you Cosmology, very much. This is how far we have gotten with Cosmology from this experiment.

[It follows a short series of charts with some tentative results not yet fully analyzed, which have been ommitted for publication in the present text, because the data analysis should improve greatly within a year or so, when the analysis of interplanetary dust is completed. It was pointed out that some of the data show that our galaxy itself is not round. The image is not symmetrical. The brightness is greater on the left side than it is on the right, and the galaxy seems to be a little bit thicker on the left than it is on the right. This could be interpreted as due to a bar of stars across the center of our galaxy. So the part that is closer to us on the left hand side is a little taller and a little brighter than the part on the right, which is further away. So it can be deduced that our galaxy is not circular, and also that our galaxy is not flat.]

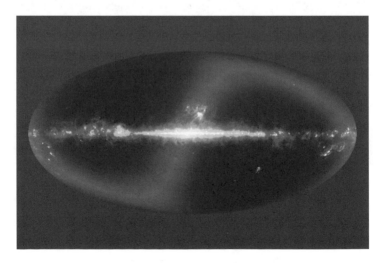

Fig. 7.43. DIRBE 678 composite.

Let me conclude by saying that you yourself can obtain our COBE skymaps and do your calculations on them (Fig. 7.44). We have just released numerical maps and they are now available, over, the computer network or by writing to us (see chart). You can write to the address in the chart or telephone, or write to me, or you can ask me questions. Next year we will be releasing more complete data sets. About a year after that we should have almost all of the data ready for general distribution.

DATA RELEASE 6/93

DIRBE GALACTIC PLANE, 10 BANDS

FIRAS GALACTIC PLANE, 105-500 μm

DMR 1YR MAPS, ALL SKY

6/94 – ALL SKY, ALL BAND, WITH POLARIZATION, TIME DEPENDENCE, ETC; DMR 2YR MAP

6/95 – IMPROVED CALIBRATION; WARN DIRBE; DMR 4YR MAP

Fig. 7.44. Do-it-yourself COBE.

To conclude I would like to talk a little bit about what comes after COBE (Fig. 7.45). We are going to turn off the COBE satellite this December because we got the data we wanted to get. We still have experiments to look for the anisotropy at the most interesting scale currently, 1° or 1/2° in size. People are doing balloon experiments, going to the Antarctica, making interferometers to use on the ground, and even thinking about sending additional space probes out to get away from the interference from the Earth atmosphere.

TURN OFF COBE 12/93

ANISOTROPY – 1/2° SCALE

    BALLOONS; LONG DURATION

    ANTARCTICA

    INTERFEROMETERS

    LZ SPACE MISSION

LONG λ SPECTRUM

    "CRATE" SMEX 1-8 cm

    IROSSE: IR OUTER SOLAR SYSTEM EXPLORER OUTSIDE ZODIACAL DUST

Fig. 7.45. Future.

Our group also proposed a long wavelength spectrum experiment, which would be a satellite experiment, what NASA is calling a Small Explorer Program. We have not yet heard whether that experiment would be approved, but I think it is an excellent idea.

Finally we know that it would be a good idea to send a probe outside the solar system to look for the IR background radiation outside the interference from the interplanetary dust.

Chapter 8

COBE Observations of the Early Universe, by George F. Smoot

The Cosmic Background Explorer (*COBE*) satellite was developed to measure the diffuse infrared and microwave radiation from the early universe, to the limits set by the astrophysical foregrounds. *COBE* has three instruments: The Differential Microwave Radiometer (DMR) maps the cosmic radiation precisely. The Far Infrared Absolute Spectrophotometer (FIRAS) compares the cosmic background radiation spectrum with that of a precise black-body. The Diffuse Infrared Background Experiment (DIRBE) searches for the accumulated light of primeval galaxies and stars. FIRAS has measured the cosmic background radiation spectrum nearly 1000 times more precisely than previous observations. The DMR discovered the cosmic background to be anisotropic at a level of a part in 10^5. The results strongly constrain cosmological models and the formation of structure in the present universe. The results from *COBE* are in good quantitative agreement with the calculations of the hot Big Bang model. The DMR data are in agreement with inflationary model predictions.

8.1. Introduction

In hot Big Bang cosmology the cosmic background radiation is the remnant of the primeval fireball beginning about 15 billion years ago, and is our best probe for examining the early universe. Its blackbody nature is a stringent test of the Big Bang theory which predicts its spectrum will be quite close to that of a blackbody. Its angular distribution contains information on the isotropy and homogeneity of the early universe. This information provides constraints on both the dynamics and geometry of the universe and on the spectrum of primordial density fluctuations which grew under the influence of gravity to present day structures.

According to Big Bang theory the universe is expanding from an original primeval fireball — a near thermal equilibrium state of very high density and temperature. As the universe expanded it cooled, and when it reached a temperature of about 3000 K the primeval plasma coalesced to neutral atomic hydrogen and helium. At that time, about 300,000 years after the Big Bang, the cosmic background radiation was free to move through the universe without scattering by free electrons. As a result the cosmic background photons are undisturbed, except for the universal expansion, since that epoch. If the primeval universe was slightly inhomogeneous, the cosmic background radiation would be slightly anisotropic.

The decoupling of light freed ordinary, baryonic matter from the radiation drag that prevented gravity from pulling it together into structures such as galaxies, clusters of galaxies, or even larger structures. An object 300,000 light years across at that time was comparable in dimension to the distance that light had traveled in the age of the universe then. Such an object at the surface of the last scattering would subtend an angle on the sky of about 1°. The Differential Microwave Radiometer (DMR) maps pixels of 2.6° so that all of the structures observed are primordial in that they could not have been formed by processes operating at the speed of light or slower in the time available to that point.

Galaxies and stars formed from these primeval inhomogeneities, but the details are unknown. The galaxies and stars continue to evolve, and their luminosities, colours, morphology and numbers are different than they were when the universe was only half its present size. The light from these first objects may have been very intense, in which case the present universe should be filled with a diffuse infrared background radiation field.

8.2. COBE Mission

The original motivations for the COBE mission focused upon cosmology. In 1974 three groups submitted to NASA proposals for instruments to make cosmological observations. From these three groups six people (Sam Gulkis, Mike Hauser, John Mather, George Smoot, Rainer Weiss, and Dave Wilkinson) were appointed to a scientific working group to study a cosmology mission concept. In the intervening years these original six people were augmented by the appointment to the COBE Science Working Group of additional members (Chuck Bennett, Nancy Boggess, Ed Cheng, Eli Dwek, Mike Janssen, Tom Kelsall, Phil Lubin, Stephan Meyer, Harvey Moseley, Tom Murdock, Rick Shafer, Bob Silverberg, and Ned Wright). Additional scientists worked on each of the instruments, and many managers, engineers, and others worked on the mission. It was through the effort of these people that the COBE mission concept was refined and realized and produced its results.

The main objectives of the COBE mission are to measure the cosmic microwave background spectrum and anisotropy, and to measure the general diffuse infrared background from early generations of objects. Secondary objectives include determination of the radiation from our local environment, such as interstellar and interplanetary dust, interstellar electrons, and starlight. The COBE instruments are sufficiently sensitive and accurate that these local sources limit the accuracy of the cosmological measurements.

The COBE team soon realized the need for a careful and well integrated mission. A successful strategy required that the mission not only must make the observations but also must provide a nearly ideal environment in terms of stability and shielding of the instruments. Thus, all instruments are protected by a conical shield which is coplanar with their apertures. The spacecraft, orbit, and operations were designed to minimize interference and variations from the environment.

The three scientific instruments selected to be on COBE are:

* the Differential Microwave Radiometer (DMR) to map the full sky at wavelengths of 9.5, 5.7, and 3.3 mm (frequencies of 31.5, 53, and

90 GHz) with a 7° resolution and a sensitivity of the cosmic background intensity better than one part in 100,000.

- the Far Infrared Absolute Spectrophotometer (FIRAS) to measure the sky spectrum over the wavelength range from 100 μm to 1 cm wavelength with a 7° resolution.
- the Diffuse Infrared Background Experiment (DIRBE) to map and measure the sky spectrum over the wavelength range from 1 to 300 μm with a 0.7° resolution in 10 broad frequency bands.

Overviews of *COBE* and its early results have been given previously (e.g. Boggess *et al.*, 1992) so that only a brief summary and up date appear here.

COBE was launched at dawn (14:34 UT) on 18 November 1989 from Vandenberg AFB, California into a 900 km altitude circular orbit, which is inclined 99° to the equator and oriented approximately perpendicular to the Sun line. The spacecraft is oriented away from the Earth and nearly perpendicular to the Sun. It spins about its symmetry axis at about 0.815 rpm. The spin period is about 73 seconds.

The combination of 99° inclination and 900 km altitude results in a precession rate (1° per day) of the orbit due to the Earth's quadrupole moment (equatorial bulge due to Earth's rotation). This precession keeps the satellite orbit plane approximately perpendicular to the sun direction as the Earth orbits the sun. *COBE* was launched toward the South in the morning so that the orbit normal pointing toward the Sun is at declination -9°. At the June solstice, when the declination of the Sun is +23.44°, there is a 32.4° deviation from the ideal situation where the Earth-Sun line is exactly perpendicular to the orbit. Since the horizon is depressed by 29° for a 900 km altitude, there are eclipses when *COBE* passes over the South pole for a period of two months centered on the June solstice. During the same season at the North pole the angle between the Sun and anti-Sunward limb of the Earth is only 177°. It is thus not possible to keep both Sunlight and Earthshine out of the shaded cavity around the dewar and instruments. Since the Sun produces a much greater energy input than a thin crescent of Earth limb, the satellite is oriented to keep the Sun below the plane of the shade.

The pointing of COBE is normally set to maintain the angle between the spin axis and the Sun at 92° to 94°. This leaves the Sun 2° to 4° below the plane of the shield. During eclipse seasons this margin is shaved to slightly less than 2° to reduce the amount of Earthshine into the shaded cavity where the instruments sit. The spin axis is aligned so the instruments look away from the Earth center. Since the Sun can be up to 32.4° from the orbit normal and the Sun is kept 2° to 4° below the plane of the shade, the spin axis can be up to 36° away from the zenith perpendicular to the orbit plane.

The angle of the spin axis in the orbit plane is rotated at a constant rate of one cycle per orbit for ease of control. This keeps the spin axis pointing near the zenith but it oscillates by as much as ±6° relative to the spin-Sun plane with a period of 2 cycles per orbit. The spin axis (pitch) is biased backward by 6° to keep the residual 900 km atmosphere from impinging on the cryogenic optics as a result of COBE's orbital motion. The DMR and DIRBE instruments have fields-of-view located 30° away from the spin axis. As a result they scan cycloidal paths through the sky, that cover the band-shaped region from 64° to 124° from the Sun. This band contains nearly 50% of the sky, and it is completely covered in a day by the DMR and DIRBE. This rapid coverage of the sky reduces the effect of drifts in the instruments. Since FIRAS coadds interferograms over a collection period of about 30 to 60 seconds, slow scanning of the sky is essential. Thus the FIRAS looks directly along the satellite spin axis for slower sky coverage at the satellite orbital of 3.6° per minute.

Over the course of a year the Earth moves around the sun, and the scan paths of the instruments follow this motion. The leading edge of the wide scanned band for the DIRBE and DMR caught up to the trailing edge of the band at start of mission after 4 months, leading to 100% sky coverage. For the FIRAS total sky coverage would in principle be obtained after 6 months, but because of gaps in the observations caused by calibration runs and lunar interference, the actual sky coverage for FIRAS, defined as having the center of the beam within a pixel, is about 90%. The coverage gaps are sufficiently narrow, however, that every direction on the sky was covered by at least the half power contour of the beam.

The COBE attitude is determined from sun and Earth sensors and gyroscope readings and is corrected using infrared star sightings by DIRBE. This method requires 3 working gyros that are not co-planar. On 10 July 1993, COBE suffered its fourth gyro failure of its six gyros, which will degrade the accuracy of the attitude solutions by about a factor of 5, to a 1σ accuracy of about 0.1°. (One of the failed gyros is noisy but operable.) This accuracy will be sufficient for the DMR experiment with its 7° beam, provided that the errors are random and not systematic. After the gyro failure the pitch was changed from 6° to 3° tilt backwards (with the same ±6° variation) and the roll was set to 2°.

The FIRAS and DIRBE instruments were cooled to about 1.5 K for 10 months inside a 650 litre liquid-helium dewar. After the helium was exhausted their detectors rose to about 60 K so that only the highest frequency bands of DIRBE continued to gather useful data. The 53 and 90 GHz microwave radiometers are radiatively cooled to 140 K, while the 31.5 GHz receiver is heated to 290 K. Thus the DMR continues to work nominally. Plans are to operate COBE until December 1993, which would complete four years of observations with the DMR and the short wavelength channels of the DIRBE.

8.3. DMR Instrument

The Differential Microwave Radiometer (DMR) instrument (Smoot *et al.*, 1990) aboard NASA's Cosmic Background Explorer (COBE) satellite was designed to provide precise maps of the microwave sky on large angular scales limited only by instrument sensitivity and observation time — ultimately perhaps by the foreground galactic emission. The DMR consists of six differential microwave radiometers, two independent radiometers at each of three frequencies: 31.5, 53, and 90 GHz (wavelengths 9.5, 5.7, and 3.3 mm). These frequencies encompass a window in which the cosmic background radiation dominates foreground galactic emission by a factor of roughly 1000. The multiple frequencies combined with data from other experiments allow subtraction of galactic emission using its spectral signature, yielding maps of the cosmic background radiation temperature.

Each radiometer has two antennas whose fields of view are separated by a 60° angle that is bisected by the COBE spin axis. Each beam has a 7° FWHM (full width half maximum). A ferrite waveguide switch connects the receiver input to one antenna and then the other at a rate of 100 Hz. The input signal then goes through a mixer, an IF amplifier and a video detector. The output of the video detector is demodulated by a lock-in amplifier synchronized to the input switch. The difference in antenna input power signal that results is boxcar integrated for 0.5 seconds and telemetered to the ground. Each radiometer has two channels: A and B. The 31.5 GHz radiometer channels share a single pair of antennas, taking as input the two opposite senses of circular polarization. Each 53 and 90 GHz radiometer channel has it own pair of antennas (4 total), and all observe the same sense of linear polarization.

Each radiometer measures the difference in microwave power between two regions of the sky separated by 60°. The combined motions of spacecraft spin (73 second period), orbit (103 minute period), and orbital precession (1° per day) compare each sky position to all others through a massively redundant set of all possible difference measurements spaced 60° apart.

A software analysis system that receives data telemetered from the COBE satellite determines the instrument calibration, and inverts the difference measurements to map the microwave sky in each channel. A significant step in the DMR data analysis is to construct a map of the sky using the 6×10^7 differences that are collected each year. We analyze the data using 6,144 pixels to cover the sky. These are of approximately equal area and arranged in a square grid of 32×32 pixels on each of the 6 faces of a cube. Within each pixel we assume that the temperature is constant. (More recently we have been using double resolution along the galactic plane region where there is sufficient signal-to-noise ratio to warrant it. This does not change this discussion significantly as for most analyses we exclude the galactic plane region or reduce the resolution.) The map-making algorithm minimizes the chi-squared statistic between the map pixel values and the difference observations. This is a least-squares problem with 6×10^7 equations in 6,144 variables. This system of equations reduces to a 6,144 by 6,144 matrix, A_{ij}, and a linear set of normal equations: $\sum_i A_{ij} T_i = Obs_j$. The DMR maps are found by solving these normal equations. Fortunately, this matrix A_{ij} is sparse and symmetric. Each pixel value, T_i is connected only to all pixels 60° away. Thus, only 9×10^5 elements are non-zero. A typical pixel has 10,000 to 20,000 observations per year. It is connected to a typical pixel 60° away by about 100 observations.

Although the experiment has been designed to minimize or avoid sources of systematic uncertainty, both the instrument and the data processing can potentially introduce systematic effects correlated with antenna pointing, which would create or mask features in the final sky maps.

8.4. DMR Limits on Potential Systematica

The sky maps may in principle contain contamination from local sources or artifacts from the data reduction process. The data reduction and analysis process must distinguish cosmological signals from a variety of potential systematic effects. These can be categorized into several broad classes.

The most obvious source of non-cosmological signals is the presence in the sky of foreground microwave sources. These include thermal emission from the COBE spacecraft itself, from the Earth, Moon, and Sun, and from other celestial objects. Non-thermal radio-frequency interference (RFI) must also be considered, both from ground stations and from geosynchronous satellites. Although the DMR instrument is largely shielded from such sources, their residual or intermittent effect must be considered.

A second class of potential systematics may result from the changing orbital environment on the instrument. Various instruments components perform slightly differently with changes in temperature, voltage, and local magnetic field, each of which is modulated by the COBE orbit and rotation. Longer-term drifts can also affect the data. Finally, the data reduction process may introduce or mask features in the data. The DMR data are differential; the map algorithm is subject to concerns of both coverage (closure) and solution stability. Other features of the data reduction, particularly the calibration and baseline subtraction, are also a source of potential artifacts.

All potential sources of systematic error must be identified and their effects measured or limited before maps with reliable uncertainties can be produced. A variety of techniques exist to identify potential systematics and place limits on their effects. For the case of known celestial sources or geosynchronous RFI, the mapping program can produce a sky map in appropriate object-centered coordinates. The contribution of the source at a given distance from beam centre can be read directly from the maps to the noise limit. Spike detection and direct inspection of the data during satellite telemetry limit the effects of asynchronous or intermittent RFI. Each real and potential systematic effect has been treated in more than one way and the results summarized in Kogut *et al.* (1992).

In addition to the COBE science working group who are coexperimenters on all three instruments, the DMR team also currently includes J. Aymon, A. Banday, G. De Amici, G. Hinshaw, A. Kogut, E. Kaita, C. Lineweaver, K. Loewenstein, P.D. Jackson, P. Keegstra, R. Kummerer, V. Kumar, J. Santana, and L. Tenorio. It is through these people and their predecessors and the others that worked on COBE that the DMR has been able to achieve a sensitivity on the order of $\delta T/T \sim 10^{-6}$.

8.5. DMR Observations

The first year data have been processed to produce maps of the microwave sky for each of the six DMR channels. The independent maps at each frequency enable celestial signals to be distinguished from noise or spurious features — a celestial source will appear at identical amplitude in both maps. The three frequencies allow separation of cosmological signals from local (galactic) foregrounds based on spectral signatures. The maps have been corrected to solar-system barycentre and do not include data taken with an antenna closer than 25° to the Moon. All six maps clearly show the dipole anisotropy and galactic emission. The dipole appears at similar amplitudes in all maps while galactic emission decreases sharply at higher frequencies, in accord with the expected spectral behaviour. Motion with velocity $\beta = v/c$ relative to an isotropic radiation field of temperature T_0 produces a Doppler-shifted temperature

$$T(\theta) = T_0 (1 - \beta^2)^{1/2} / (1 - \beta \cos\theta)$$
$$\approx T_0 \left(1 + \beta\cos\theta + (\beta^2/2)\cos(2\theta) + O(\beta^3)\right)$$

$$(8.1)$$

The first term is the monopole cosmic background radiation (CBR) temperature without a Doppler shift. The term proportional to β is a dipole, varying as the cosine of the angle between the velocity and the direction of observation. The term proportional to β^2 is a quadrupole, varying with the cosine of twice the angle with amplitude reduced by $1/2\beta$ from the dipole amplitude.

The DMR maps clearly show a dipole distribution consistent with a Doppler-shifted thermal spectrum. The DMR finds the same thermodynamic amplitude for all three frequencies. Using the DMR direction and FIR AS spectrum one finds the difference in spectra taken near the hot and cold poles is described to better than 1% of its peak value by the difference in blackbody spectra.

The DMR maps are well-fitted by a dipole cosine dependence. There is annual modulation by the 30 km/s^{-1} Earth orbital velocity. Everything is consistent with the kinematic effect of our motion relative to the CMB rest frame and thus a Doppler shift origin.

The implied velocity for the solar-system is $\beta = 0.00123 \pm 0.00003$ towards $(\alpha, \delta) = (11.2^h \pm 0.2^h, -7° \pm 2°)$, or $(l,b) = (265° \pm 2°, 48° \pm 2°)$. The solar system velocity with respect to the local standard of rest is about 20 km/s toward $(l, b) = (57°, 23°)$, while galactic rotation moves the the local standard of rest at 220 km/s toward $(90°, 0°)$ (Fich, Blitz, and Stark 1989). The DMR results thus imply a peculiar velocity for the Galaxy of $v_g = 550 \pm 10$ km/s in the direction $(266° \pm 2°, 30° \pm 2°)$ consistent with an independent determinations of the velocity of the local group, $v_{lg} = 507 \pm 10$ km/s toward $(264° \pm 2°, 31° \pm 2°)$ (Kerr, F.J., and Lyndon-Bell, D., 1990, Yahil, A., Tamman, A., and Sandage, 1977).

A summation of gravitational attractors in the local area (red shift < 6000 km/s) find reasonable agreement between a gravitationally induced velocity and the CMB dipole magnitude and direction (e.g. Bertschinger *et al.*, 1990).

For the DMR maps with the best-fitted Doppler effect removed from the data, the only high signal-to-noise ratio large-scale feature remaining is galactic emission, confined to the plane of the galaxy. This emission is consistent with emission from electrons (synchrotron and HII) and dust within the galaxy. When one analyzes the maps more thoroughly other features become apparent. A natural question is: are the features galactic? We have studied galactic emission and the features and conclude that the effect is not due to galactic emission as we understand it at the present time (Bennett *et al.*, 1992). There might be classes of emitters which have not yet been detected or have a much different spatial distribution of emission than observed at other frequencies or a net conspiracy of the three major sources of galactic emission, synchrotron, free-free, and dust, to alias a thermal spectrum. This is fairly unlikely.

Extragalactic radio sources are expected to produce variations in the DMR beam of about 1 to 2 μK. Less is expected from infrared sources. Again one might envision a new class of emitters or effects.

There is evidence of other features. These include a quadrupole anisotropy and anisotropy at higher order spherical harmonics. We have made a series of spherical harmonic fits to the data, excluding data within several ranges of galactic latitude. The quadrupole and higher-order terms are limited to $\delta T/T < 10^{-5}$. However, there are statistically significant features at slightly lower levels. The best-fitted amplitude for the observed rms quadrupole amplitude was $13 \pm 4/\mu$K corresponding to $\delta T/T \approx 5 \times 10^{-6}$. The rms fluctuations on a $10°$ angular scale was found to be 30 ± 5 μK corresponding to $\delta T/T \approx 1.1 \times 10^{-5}$.

One can continue the fitting to spherical harmonic components

$$T(\theta,\phi) = \sum_{l} \sum_{m=-l}^{l} a_{lm} Y_{lm}(\theta,\phi) \qquad (8.2)$$

where the a_{lm} are the spherical harmonic coefficients. (We actually calculate the b_{lm} using modified spherical harmonics in order to use only real numbers but the results are the same (Smoot *et al.*, 1991).) One can compute the rotationally-invariant multipole moment power, the ΔT_l^2, according to the relation

$$\Delta T_l^2 = \frac{1}{4\pi} \sum_m |a_{lm}|^2 = \frac{1}{4\pi} \sum_m |b_{lm}|^2 \qquad (8.3)$$

The dipole amplitude is down by about one part in a thousand or a million in power. The quadrupole amplitude is down by about another factor of 200 from the dipole or about 4×10^{-10} in power from the monopole. The higher order terms are at roughly this same level. A major problem is that the non-uniform sky coverage, particularly the galactic latitude cuts made to exclude significant galactic emission, result in the spherical harmonics no longer being orthogonal. As a result, alternate tacks must be taken to account for this correlation. The power spectrum for the COBE DMR maps can be determined by three methods: fitting masked spherical harmonics to the maps and finding the quadrature sum of the power at each l, fitting directly to the map for the power at each l, or by fitting the correlation function.

The correlation function, $C(\alpha) = \langle T(n_1)T(n_2) \rangle$, is the average product of temperatures separated by angle a. It is calculated for each map or in cross correlation between maps by multiplying all possible pixel pairs outside a galactic latitude cut, typically 20°, and averaging the results into 2?6 bins after removing the least squares fitted mean, dipole, and quadrupole. The correlation function is related to the sky temperature power spectrum,

$$C(\alpha) = \sum_{l>2} \Delta T_l^2 W(l)^2 P_l(\cos\alpha) \qquad (8.4)$$

where P_l are the Legendre polynomials, and $W(l)$ is the beam weighting or filter function. A 3.2° rms Gaussian beam gives weights $W(l) = \exp[-1/2(l(l + 1)/17.8^2)]$. However, one should use the weights computed from the actual DMR beam patterns and given in Wright *et al.* (1994a, b; or in more detail by Kneissl & Smoot 1993). Though the beam is reasonably well matched by a Gaussian beam shape, the difference in $W(l)^2$ can be up to 20% for values of l between 8 and 12.

If one has a model of the primordial (or surface of last scattering) fluctuations which predicts the power spectrum of fluctuations, then one can readily calculate the model correlation function $C(\alpha)$, and compare it with the DMR correlation function. One particularly interesting version is the scale-free power spectrum -a simple power law in physical size or angle.

A major finding is that the fluctuations are well-fitted by a normally-distributed, scale-invariant spectrum. With the dipole and quadrupole removed from the data, the best-fit scale-free spectrum had an equivalent rms quadrupole amplitude of 16 ± 4 µK and a power law index of 1.1 ± 0.6, compared with 1 for a scale-invariant spectrum. Similarly, a search for non-Gaussian fluctuations on the sky showed no features to $\delta T/T < 8 \times 10^{-5}$. The results are insensitive to the precise cut in galactic latitude and are consistent with the expected Gaussian instrument noise and scale invariant fluctuations. The reported uncertainties are at the 95% confidence level unless otherwise stated, and include the effects of systematics as listed in Table 8.1. Smoot *et al.* (1992), Bennett *et al.* (1992), Wright *et al.* (1992), and Kogut *et al.* (1992) reported the basic DMR result — the discovery of an intrinsic anisotropy of the microwave background beyond the dipole anisotropy.

The power spectrum remains of great interest as it is a convenient way to compare observational results among the various experiments as well as to theoretical predictions. As noted, non-uniform sky coverage creates non-orthogonality of the spherical harmonics and some ambiguity of results. One approach is to create a set of orthogonal basis functions for the experimental sky coverage and fit the data, and then convert back to the spherical harmonic power. Wright (1993) has computed the power spectrum of the DMR maps using the Hauser–Peebles method to allow for incomplete sky coverage caused by masking out the galactic plane region. Figure 8.1 shows the power in each multipole normalized to a pure $n = 1$ (Harrison–Zeldovich) power law with $\Omega_{rms-PS} = 17$ µK. Also shown are results from various small and medium scale ΔT experiments. These results are similar to what one obtains fitting to the correlation function and directly to the power in the maps.

The COBE DMR measurement of CBR anisotropics provides significant new information on the formation of large-scale structure, as well as the early universe. The dipole anisotropy provides a precise measure of the Earth's peculiar velocity with respect to the co-moving frame. The results confirm some of the major predictions of inflation, namely the prediction of normally-distributed, nearly scale-invariant fluctuations (Bardeen *et al.*, 1982; Guth & Pi 1982; Hawking 1982 and Starobinskii 1982).

Since we do not believe the observation is due to an instrumental or experimental effect, a local source, or our Galaxy or other galaxies, we make the assumption that the large-scale structure observed is primarily due to effects at the surface of last scattering, i.e. about one optical depth for Thomson scattering.

Table 8.1. Major classes of potential systematic effects.

Class Effect	Typical 95% Confidence Limit (μK)
Foreground Emission	
Earth	0.7
Moon	0.1
Sun	0.4
Planets (Jupiter)	0.2
RFI	0.0
COBE Shield	0
Zodiacal Dust	0
Environmental Susceptibilities	
Magnetic	1.9
Thermal	1.5
Voltage	0.1
COBE Cross-talk	$\ll 1$
Miscellaneous	5
Processing & Software Artifacts	
Absolute Calibration	2
Calibration Drifts	0.3
Antenna Pointing	0.2
Solution Stability	0.1
Noise Properties	0.1
Confusing Sources	
Galactic Emission	3
Extragalactic Sources	1–2

COBE Observations of the Early Universe

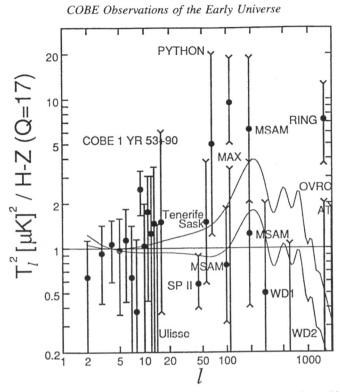

Fig. 8.1. The angular power spectrum of the CMBR, normalized to a $<Q^2_{rms}>^{0.5} = 17 \mu K$ Harrison–Zeldovich model. Data from l = 2 to 18 are from COBE, except for three points from the confirming MIT (FIRS) experiment (Ganga *et al.*, 1993). Other data and upper limits are from left to right: Ulisse (de Bernardis *et al.*, 1992), Tenerife (Watson *et al.*, 1992), SP H (Schuster *et al.*), PYTHON (Dragovan, 1993), MS AM (Cheng *et al.*, 1993), two points from MAX (Meinhold *et al.*, 1993; Gundersen *et al.*, 1993), MSAM again, and OVRO (Readhead *et al.*, 1989).

8.6. DMR Result Discussion

The origin of large-scale structure in the universe is one of the most important issues in cosmology. Currently the leading models for structure formation postulate gravitational instability operating upon a primordial power spectrum of density fluctuations. The inflationary model of the early universe produces primordial density fluctuations with a nearly scale-invariant spectrum, which suggests a viable mechanism for structure formation. Structure forms as the result of gravitational amplification of initially small perturbations in the primordial mass-energy distribution, and non-baryonic dark matter seems necessary to provide sufficient growth of these perturbations. The determination of the nature of the initial density fluctuations then becomes an important constraint on cosmological models.

The discovery of the anisotropy in the cosmic microwave background radiation by the COBE DMR instrument and the recent confirmation (Ganga *et al.*, 1993) mark a new era in cosmology and the beginning of investigations of these primordial fluctuations. The frequency independence of the anisotropy is a strong argument that the anisotropy is in the cosmic microwave background radiation and not due to foreground galactic emission or extragalactic sources. The confirming 'MIT' balloon-borne bolometer observations (Ganga *et al.*, 1993) have an effective frequency of about 170 GHz, which makes the argument stronger. The typical fluctuation amplitude is roughly 30 μK or $\Delta T/T \sim 10^{-5}$ on a scale of $10°$. The data appear consistent with a scale-invariant power spectrum with an uncertainty of ± 0.6 in the exponent of the power law. The amplitude and spectrum are consistent with that expected for gravitational instability models involving nonbaryonic dark matter, and perhaps consistent with the measured large scale velocity flows. The angular power spectrum of the DMR maps is estimated by several methods, and is consistent with a near scale-invariant power spectrum drawn from a Gaussian distribution.

The primary interpretation of the temperature anisotropics observed by the DMR is in terms of metric perturbations on the surface of last scattering. Sachs and Wolfe (1967) derived an expression relating the metric pertubations to temperature anisotropics. Often, when considering density perturbations, the Sachs-Wolfe relation is simplified to its dominant term: $\delta T/T = \Delta\phi/3c^2$ where $\Delta\phi$ is the gravitational potential fluctuation at the surface of last scattering.

The ratio of the gravitational potential fluctuations seen by the DMR through the Sachs-Wolfe effect to the gravitational potentials inferred from the bulk flows of galaxies using the POTENT method (Bertschinger *et al.* (1990)), shows that the Harrison–Zeldovich spectrum has the correct slope to connect the DMR data at 3×10^5 km/sec scales to the bulk flow data at 6000 km/sec scale. This comparison of the larger than horizon-scale perturbations seen by COBE and the "large-scale" (60/h Mpc, where h = H_0/(100 kms^{-1} Mpc^{-1})) structure data gives best determination of the slope of the power spectrum of the primordial perturbations, even though it is model-dependent. COBE data alone covers such a small range of scales that slope determinations are very uncertain.

The DMR pixel and beam size means that all perturbations observed are larger than the horizon size at the surface of last scattering assuming that it is at red shift of ≈ 1100. For perturbations at scales smaller than the horizon at the end of radiation dominance, dynamical effects boost the expected anisotropy. At very small scales, the finite thickness of the recombination surface filters out most of the anisotropy. The two models shown in Fig. 8.1 are from Crittenden *et al.* (1993), and correspond to an n = 0.85 tilted CDM model (with H_0 = 50 kms^{-1} Mpc^{-1} and $\Omega_B h^2$ = 0.0125), and the same and correspond to an n = 0.85 tilted CDM model (with H_0 = 50 kms^{-1} Mpc^{-1} and $\Omega_B h^2$ = 0.0125), and the same model with the n = 0.96 expected from chaotic ϕ^4 inflation. Clearly much better measurements of small scale ΔT are needed to have any hope of measuring the tilt. The POTENT (Bertschinger *et al.*) bulk flow, if translated into a ΔT_l^2 in Fig. 8.1, would correspond to a point on the higher model curve at $l \approx 100$. The Lauer & Postman (1992) bulk flow would correspond to a ΔT_l^2 that is ≈ 40 times higher than the higher model at $l \approx 40$.

Wright *et al.* (1992) discussed the implications of the COBE DMR data for models of structure formation, and selected 4 models from the large collection in Holtzman (1989) for detailed discussion: a "CDM" model, with H_0 = 50 kms^{-1} Mpc^{-1}, Ω_{CDM} = 0.9, and Ω_B= 0.1; a mixed "CDM+HDM" model, with H_0 = 50 kms^{-1} Mpc^{-1}, Ω_{CDM} = 0.6, Ω_{CDM} = 0.3 (a 7 eV neutrino), and Ω_B = 0.1; an open model, with H_0 = 100 kms^{-1} Mpc^{-1}, Ω_{CDM} = 0.18, and Ω_B = 0.02; and a vacuum dominated model, with H_0 = 100 kms^{-1} Mpc^{-1}, Ω_{CDM} = 0.18, Ω_B = 0.02, and Ω_{vac} = 0.8. The vacuum dominated model and especially the open model have potential perturbations now that are too small to explain the Bertschinger *et al.* bulk flow observation. Note that all of these models are normalized to the COBE anisotropy at large scales at $z = 10^3$, but in the open and vacuum-dominated models the potential perturbations go down even at very large scales which are not affected by non-linearities or the transfer function. The small-scale anisotropics predicted by these models are quite close to the upper limits, and in the case of the 1.5° beam South Pole to the upper limits, and in the case of the 1.5° beam South Pole experiment they exceed the current data. Since the DMR data are effectively centered at $l_{eff} = 4$ while the Schuster *et al.* datum is at $l_{eff} = 44$, a tilted model which replaces the Harrison–Zeldovich $n = 1$ in $P(k) \propto k^n$ with $n \approx 0.5$ would eliminate this discrepancy. But such a low n would destroy the agreement between the bulk flow velocities and the COBE amplitude. In fact the bulk flows are inconsistent with the Schuster *et al.* data (Gorski 1992), but consistent with the Saskatoon data (Wollack *et al.*, 1994).

The DMR maps show statistically significant correlated structure with good agreement between the DMR correlation function and the Harrison–Zeldovich model shown by Smoot *et al.* (1992). Thus, while supporting inflation, the COBE results do not prove inflation.

The inflationary scenario also predicts that the temperature fluctuations should have a Gaussian distribution. This can be tested by looking at higher-order moments of the maps such as three-point and higher correlation functions, or by directly studying the probability distributions. Another approach is to use topological measures of the map structures. This work is currently in progress; however, at this stage the random measurement noise from the radiometers is still sufficient to interfere with these studies to a great extent. A preliminary analysis finds that the Gaussian model is consistent with the observations, that is, the data are compatible with the sum of instrument noise plus a scale-invariant, Gaussianly distributed spectrum of fluctuation amplitudes. There is no evidence of non-Gaussian behaviour. This might be considered a blow against "defect" versus gravitational instability models of structure formation, except calculations by Bennett & Rhie (1993) show that models which predict non-Gaussian temperature fluctuations, such as cosmic strings or global monopoles, can also match the observed data. This convergence toward Gaussian probability distributions is caused by the 10° effective resolution of the smoothed maps, which averages many small patches together. Thus, cosmic strings, which predict dramatically non-Gaussian behaviour (sharp edges) on small scales, give almost Gaussian results after smoothing.

The horizon due to expansion of the universe limits the region over which we can observe these primordial fluctuations. For the very largest scales only a small number of fluctuations will be present inside our horizon, creating error due to our cosmic sampling variance. If the fluctuations are drawn from a Gaussian distribution, the cosmic variance will limit the DMR's ability to determine the mean cosmic fluctuation amplitude to about 10%.

Since the DMR detection was announced, several medium and smaller angular scale experiments have new results. These include: the UCSB South Pole experiment (Gaier *et al.*, 1992, Schuster *et al.*, 1993), the Princeton–Saskatoon experiment (Wollack *et al.*, 1994), the Tenerife collaboration (Hancock et al. 1994; Watson *et al.*, 1992), the CARA South Pole experiments — Python (Dragovan *et al.*, 1993) and White Dish (Peterson *et al.*, 1993), the Center for Particle Astrophysics MAX balloon-borne experiment (Devlin *et al.*, 1994; Gunderson *et al.*, 1993; Meinhold *et al.*, 1993), the MS AM balloon-borne experiment (Cheng *et al.*, 1993), the Owens Valley Radio-Astronomy Observatory (OVRO) (Myers *et al.*, 1993), the Roma balloon-borne experiment ULISSE (de Bernardis *et al.*, 1992), and the Australia Telescope (Subrahmayan *et al.*, 1993). In general these experiments report fluctuations at the 10^{-5} level. These measurements could in principle distinguish among models of structure formation, shed information on the nature of the dark matter, and probe the existence of cosmological gravity waves (Crittenden *et al.*, 1993; Smoot & Steinhardt 1993). However, the current results vary at the factor of two to three level, which is just what is needed to make the distinctions. There is perhaps evidence that the data are not self-consistent. In part, one can assume that some discrepancy is due to experimental error, to the limited region of the sky sampled by these experiments, and to the potential confusion by galactic emission and extragalactic point sources. The community eagerly awaits refined observations of the power spectrum. In order to distinguish the various models, one needs to measure the large angular-scale power spectrum as accurately as possible, given cosmic variance, and utilize that as a normalization of the primordial spectrum for comparison with other observations and theoretical models.

8.7. DIRBE Measurements

The COBE Diffuse Infrared Background experiment (DIRBE) is designed to map the infrared and far infrared sky, and to search for a cosmic infrared background (CIB). Models of the CIB radiation take visible light from distant, unresolved galaxies, and Doppler shift it into the near infrared bands (Partridge & Peebles 1967) or absorb the starlight on dust, reradiate it in the far-infrared, and Doppler shift it into the sub-millimeter band (Stecker *et al.*, 1977). These models predict fluxes that

Table 8.2. Brightness and Planck brightness temperature of the diffuse infrared sky.

Reference	λ (μm)	λI_λ (10^{-7} W m^{-2} sr^{-1})	$T_B(\lambda)$ K
DIRBE	1.2	8.3 ± 3.3	351
(South Ecliptic Pole)	2.3	3.5 ± 1.4	216
	3.4	1.5 ± 0.6	140
	4.9	3.7 ± 1.5	109
	12.0	29.0 ± 12	56
	22.0	21.0 ± 8	34
	55.0	2.3 ± 1	15
	96.0	1.2 ± 0.5	9.2
	151.0	1.3 ± 0.7	6.6
	241.0	0.7 ± 0.4	4.5

are approximately two orders of magnitude lower than the emission from our Milky Way Galaxy, which makes detection of the CIB a very difficult task, unless observations are made at wavelengths where the redshifted light carries the energy from distant galaxies into a region with little Milky Way emission. The interplanetary dust cloud, which scatters sunlight to give the zodiacal light, absorbs much more power than it scatters and reradiates this power in the mid-infrared. Thus, from 5 μm to

100 μm the CIB is overwhelmed when observing at a distance of 1 A.U. from the Sun. However, there are windows between the scattered and thermal zodiacal light at 3.5 μm and between the interstellar dust in the Milky Way and the CMBR at 200–300 μm where the CIB may be detectable.

The DIRBE instrument is an off-axis Gregorian telescope with a 19 centimeter diameter primary. The Gregorian design includes an intermediate image of the sky which is occulted by a stray-light stop between the primary and the secondary. A second stray-light stop is located where the secondary forms an image of the primary mirror. Twenty-five baffle vanes in the telescope tube absorb far off-axis radiation, so scattering on the primary mirror when it is illuminated by a slight off-axis source is the main path for stray light. Great care was taken to avoid contamination and the preserve the cleanliness of the optics, so the off-axis response of the DIRBE is very small. The DIRBE beam is an $0.7° \times 0.7°$ square. A 32-Hz resonant tuning fork serves as a chopper giving a well-determined zero level by comparing the sky signal to the 2 K interior of the DIRBE. Less frequently a shutter blocks the optical path so the accuracy of the zero level can be checked. When the shutter is in the beam, its back side is a mirror which reflect light from a commandable internal reference source into the detectors, providing a gain calibration several times per orbit. Dichroics and aperture division are used to gives simultaneous measurements in 10 spectral bands in the same field of view.

The 10 bands in DIRBE are the ground-based photometry bands J[1.2 μm], K[2.3 μm], L[3.4 μm], and M[4.9 μm]; the four IRAS bands at 12, 25, 60, and 100 μm; and two long wavelength bands at 140 and 240 μm. The shortest wavelengths (JKL) bands, which observe scattered light from the interplanetary dust cloud, have 3 detectors per band with polarizers to measure linear polarization. During the 10 months of cryogenic operation, the achieved instrument rms sensitivity per field of view was $\lambda I(\lambda) = (1.0, 0.9, 0.6, 0.5, 0.3, 0.4, 0.4, 0.1, 11.0, 4.0) \times 10^{-9}$ W m^{-2} sr^{-1}, respectively for the ten wavelength bands. These noise levels are considerably smaller than the total fluxes measured by DIRBE at the South Ecliptic Pole and presented in Table 8.1. The extragalactic flux predicted by Partridge & Peebles (1967) and Stecker *et al.* (1977) is 3×10^{-8} W m^{-2} sr^{-1} at both 3.5 and 240 μm, which is 20% of the total SEP flux at 3.5 μm and 43% at 240 μm.

The final determination of the magnitude of the extragalactic background will depend on modeling and removing the zodiacal light. The DIRBE beam is aligned 30° from the spin axis of COBE, so it swings from 64° elongation to 124° elongation in every rotation of the satellite. In principle, the variation of intensity with elongation can be used to determine the magnitude of the zodiacal emission, just as the variation of atmospheric emission with secant of the zenith angle allows one to determine the atmospheric emission. However, the shape of the atmosphere is well known, while the shape of the zodiacal dust cloud is poorly determined. The zodiacal dust is diffusely distributed; but, it is certainly more complicated than a simple ellipsoidal cloud. The zodicial dust cloud and Earth orbit plane are clearly offset. There is an annual variation in the brightness of the north and south ecliptic poles at the 1% level which is anticorrelated between north and south. DIRBE and IRAS have also imaged two zodiacal dust band pairs at ecliptic latitudes of ±1.4° and ±10°. IRAS data show evidence of a third pair at ±14° and suggest a fourth pair at ±24°. Certainly a great deal of effort will be needed to remove of the zodiacal light to an accuracy of 1% of its peak emission.

After the depletion of the liquid helium, the DIRBE JKLM channels have continued to operate. The interior temperature of the Dewar appears to have stabilized at ≈55 K, which is adequate for the InSb detectors used in these bands. While the detector noise in these channels has risen due to the increased temperature, the noise while observing the sky is dominated by the confusion noise due to the multitude of faint stars. These warmer data have been used to provide star sightings for the attitude determination.

Table 8.2 shows the measured fluxes observed by DIRBE for the south ecliptic pole, one of the dimmer regions of the sky. These can be used as hard upper limits on the CIB (cosmic infrared background). For the longer wavelengths the galactic poles are slightly dimmer and lower fluxes appear. An accurate estimate of the zodiacal dust emission is necessary for a more restrictive determination of the CIB.

8.8. FIRAS Instruments Descriptions

The FIRAS instrument is designed to compare the spectrum of the cosmic microwave background radiation with that of a precise blackbody, enabling the measurement of very small deviations from a black body spectrum. It has a 7° diameter beamwidth, established by a nonimaging parabolic concentrator, which has a flared aperture to reduce diffractive side-lobe responses. The instrument is calibrated by a full beam, temperature-controlled external black body, which can be moved into the beam by command. The FIRAS is the first instrument to measure and compare the background radiation with such an accurate external full beam in-flight calibrator. The spectral resolution is obtained with a polarizing Michelson or Martin & Puplett (1970) interferometer with separated input and output beams to permit fully symmetrical differential operation. There are two symmetric inputs and outputs. One input beam views the sky or the full aperture external calibrator, while the second input beam views an internal temperature-controlled reference black body through its own parabolic concentrator. Both input concentrators and calibrators are temperature controlled and can be set by command to any temperature between 2 and 25 K. The standard operating condition was for the two concentrators and the internal reference body to be commanded to match the sky temperature, thereby yielding a nearly nulled interferogram and reducing almost all instrumental gain errors to negligible values.

The FIRAS instrument covers two frequency ranges, a low frequency channel from 1 to 20 cm^{-1} and a high frequency channel from 20 to 100 cm^{-1}. The separation of low and high frequency outputs is accomplished by a capacitive grid dichroic filter producing four channel outputs. The instrument's apodized spectral resolution is limited by the maximum stroke length to 0.2 cm^{-1} in the low frequency channels, and by the microprocessor buffer size and telemetry bandwidth to 0.8 cm^{-1} in the high frequency channels. Beam divergence within the instrument limits the spectral resolution to ~1% and shifts the effective frequency scale by 0.4%. The rms sensitivity for frequencies from 2 to 20 cm^{-1} is $vI_v < 10^{-9}$ W/m^2 sr per 7° field and per 5% spectral resolution element after one year of operation.

Each channel has a large-area (0.5 cm^2), composite bolometer that senses incident power by absorbing the radiation, converting it into heat, and then detecting the temperature rise of a substrate using a very small and sensitive silicon resistance thermometer. In order to maximize the temperature rise for a given input power, one minimizes the specific heat of the bolometer. The substrate is made of the material with the highest known Debye temperature: diamond. Diamond is quite transparent to millimeter waves, so it is coated with an absorbing layer of gold alloyed with chromium. The absorbing layer thickness is set to have a surface impedance about one-half of the 377 Ohms/D impedance of free space in order to match the impedance for efficient absorption. In order to detect the longest waves observed by FIRAS, the diameter of the octagonal diamond substrate is quite large, about 8 mm. Each detector sits behind a compound parabolic concentrator, or Winston cone, which concentrates a full π steradians of the incoming beam. As a result, the FIRAS instrument has a very large etendue of 1.5 cm^2 sr.

The external calibrator determines the accuracy of the instrument for broad band sources like the cosmic background radiation. It is a re-entrant cone shaped like a trumpet mute and made of Eccosorb CR-110 iron-loaded epoxy. The angles at the point and groove are 25°, so that a ray reaching the detector has undergone seven specular reflections from the calibrator. The calculated reflectance for this design, including diffraction and surface imperfections, is less than 10^{-4} from 2 to 20 cm^{-1}. Measurements of the reflectance of an identical calibrator in an identical antenna using coherent radiation at 1 cm^{-1} and 3 cm^{-1} frequencies confirm this calculation. The instrument is calibrated by measuring spectra with the calibrator in the sky horn while operating all other controllable sources within the instrument at a sequence of different temperatures. The overall responsivity and the emissivity of each source can be determined relative to the external calibrator by solving a set of coupled linear equations.

The first results of the FIRAS instrument (Mather *et al.*, 1990) may be summarized as follows. The intensity of the background sky radiation is consistent with a blackbody at 2.735 ± 0.06 K. Deviations from this blackbody at the spectral resolution of the instrument are less than 1% of the peak brightness. There is no evidence of a distortion of the spectrum and the measured temperature is consistent with previous reports and the UBC rocket result of Gush *et al.* (1990). Since early 1990 there have been significant improvements in the FIRAS measurements and results.

8.9. FIRAS Measurements

Because the FIRAS is a Michelson interferometer, the spectral data are obtained in the form of interferograms. A channel output is an intensity, *I*, which depends upon the differential mirror displacement, *x*, and is the Fourier transform of the difference in the input power from the two symmetric inputs and the instrument. The primary contributors are the internal calibrator on one side and the sky or the external calibrator on the other.

There are minor contributions from the reference horn that connects to the internal calibrator, the sky horn, the bolometer housing, the optical structure of the FIRAS, and the dihedral mirrors that move to provide the variation in path length difference x. Operating the FIRAS instrument near the cosmic microwave background temperature not only minimizes the gain errors but also the small differential absorption and emission of the instrument components. By varying conditions and temperatures of the various components the relative emissivities and calibration coefficients can be determined. Fixsen *et al.* (1993a) provides a comprehensive description of the calibration procedure.

After calibration the sky data were analyzed to determine $I_v(l,b)$, the intensity of the sky as a function of frequency, galactic longitude and galactic latitude. For each direction on the sky the observations are summarized as a combination of cosmic plus galactic signals:

$$I_v(l,b) = e^{-\tau_v(l,b,\infty)} \left(B_v \left(T_0 + \Delta T(l,b) \right) + \Delta I_v \right) \\ + \int e^{-\tau_v(l,b,s)} j_v(l,b,s) ds \tag{8.5}$$

where $\tau_v(l,b,s)$ is the optical depth between the Solar system and the point at distance s in the direction (l,b) at frequency v, ΔI_v is an isotropic cosmic distortion, and ΔT (l,b) is the variation of the background temperature around its mean value T_0. This equation can be simplified because the optical depth of the galactic dust emission is always small in the millimeter and sub-millimeter bands covered by FIRAS.

$$I_v(l,b) \approx B_v(T_0 + \Delta T(l,b)) + \Delta I_v + \int j_v(l,b,s)ds \qquad (8.6)$$

Even in the optically thin limit, some restrictive assumptions about the galactic emissivity j_v are needed, since the galactic intensity $\int j_v(l,b,s)ds$ is a function of three variables, just like the observed data. A simple model for the galactic emission is

$$\int j_v(l,b,s)ds = G(l,b)g(v) \qquad (8.7)$$

This model assumes that the shape of the galactic spectrum is independent of direction on the sky. It is reasonably successful except that the galactic center region is clearly hotter than the rest of the Galaxy. The application of this model proceeds in two steps. The first step assumes that the cosmic distortions vanish, and that an approximation $g_0(v)$ to the galactic spectrum is known. A least squares fit over the spectrum in each pixel then maps $\Delta T(l, b)$ and $G(l, b)$. The high frequency channel of FIRAS is used to derive $G(l, b)$ because the galactic emission is strongest there. An alternative way to derive $G(l, b)$ is to smooth the DIRBE map at 240 μm to the FIRAS 7° beam. The second step in the fitting then derives spectra associated with the main components of the millimeter wave sky: the isotropic cosmic background, the dipole anisotropy, and the galactic emission. All the pixels (except for the galactic centre region with $|b| < 20°$ and $320° < l < 40°$ are fitted at each frequency to the form

$$I_v(l,b) = I_v(v) + D(v)\cos\theta + G(l,b)g(v) \qquad (8.8)$$

The FIRAS calibration model is so complicated and the cosmic distortions, if any, are so small, that systematic uncertainties limit the accuracy of the isotropic spectrum $I_0(v)$ derived from the whole FIRAS data set. The best estimate of $I_0(v)$ comes from the last six weeks of the FIRAS operation, when a. modified observing sequence consisting of 3.5 days of observing the sky with the internal calibrator set to null the cosmic spectrum, followed by 3.5 days of observing the external calibrator with its temperature set to match the sky temperature T_0. Six cycles between sky and external calibrator were obtained before the helium ran out. By processing both the sky data and the external calibrator data through the same calibration model, and then subtracting the two spectra, it was found that nearly all of the systematic errors cancel out. The high galactic latitude sky intensity difference is then modeled as

$$I_v(\text{sky},|b|> b_c) - I_v(\text{external calibrator})$$

$$= \Delta I_v + \delta T \frac{\partial B_v}{\partial T} + Gg(v) \qquad (8.9)$$

The result of the least squares fit to minimize ΔI_v by adjusting δT and G is shown in Fig. 8.2. Since all of the data taken during the last six weeks of FIRAS operation were taken in a single scan mode, the mechanical resonance frequency of the mirror transport mechanism occurs at a fixed spectral frequency, leading to the large error bar on the point at 11 cm^{-1}. It is important to remember that any cosmic distortion with a spectral shape that matches one with the shape of the galactic emission or the dipole spectrum will be hidden in this fit. The maximum residual between 2 and 21 cm^{-1} is 1 part in 3000 of the peak of the black body, and the weighted rms residual is 1 part in 10^4 of the peak of the black body. While these residuals are small, they are nonetheless more than twice the residuals that one would expect from detector noise alone. Limits on cosmological distortions can now be set by adding terms to the above fit, finding the best fit value with minimum χ^2 statistic, and then finding the endpoints of the 95% confidence interval where $\chi^2 = \chi^2_{min} + 4$. As an example of such a fit, suppose that the sky in fact was a gray-body with an emissivity $\epsilon = 1 + e$, where e is a small parameter. Since the external calibrator is a good black body, this model predicts a residual of the form $eB_v(To)$. Because $B_v(T_0)$ peaks at a lower frequency than $\partial B_v / \partial T$, this model is sufficiently different from the dipole and galactic spectra that a tight limit on e can be set, namely $|e| < 0.00041$ (95%

confidence limit). The result of similar fits for the Compton y parameter and the chemical potential, μ, of the Bose-Einstein distribution are $|y| < 2.5 \times 10^{-5}$ and $|\mu| < 3.3 \times 10^{-4}$.

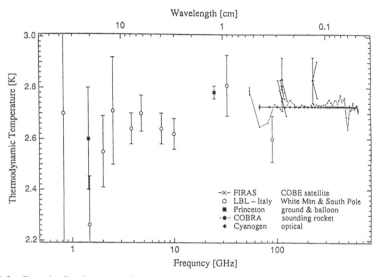

Fig. 8.2. Cosmic Background Spectrum Measurements for FIRAS and various other experiments to date — COBRA is the UBC rocket experiment (Gush *et al.*, 1990), Princeton is the balloon experiment of Johnson, Wilkinson and Staggs (1989), LBL and Italy is the White Mountain and South Pole results of my group and our Italian collaborators, Cyanogen is the interstellar cyanogen optical observations

The absolute temperature of the cosmic background, T_0, can be determined two ways using FIRAS. The first way is to use the readings of the germanium resistance thermometers in the external calibrator when the external calibrator temperature is set to match the sky. This gives $T_0 = 2.730$ K. The second way is to measure the frequency of the peak of $\partial B_\nu / \partial T$ by varying Tx a small amount around the temperature which matches the sky, and then apply the Wien displacement law to convert this frequency into a temperature. This calculation is done automatically by the calibration software, and it gives a value of 2.722 K for T_0. Three additional determinations of T_0 depend on the dipole anisotropy. For either FIRAS or DMR, the spectrum of the dipole anisotropy can be fit to the form

$$D(v) = \frac{T_0 v}{c} \frac{\partial B_v(T_0)}{\partial T} \qquad (8.10)$$

Since the velocity of the solar system with respect to the CMB is not known a priori, only the shape and not the amplitude of the dipole spectrum can be used to determine T_0. For FIRAS, this analysis gives $T_0 = 2.714 \pm 0.022$ K (Fixsen *et al.*, 1993b), while for DMR it gives $T_0 = 2.76 \pm 0.18$ K (Kogut *et al.*, 1994). The DMR data analysis keeps track of the changes in the dipole caused by the variation of the Earth's velocity around the Sun during the year. In this case the velocity v is known, so T_0 can be determined from the amplitude of the change in the dipole, giving $T_0 = 2.75 \pm 0.05$ K. The final adopted value is 2.726 ± 0.010 K (95% confidence, Mather *et al.*, 1993), which just splits the difference between the two methods based on the FIRAS spectra. The dipole-based determinations of T_0 are less precise but provide a useful confirmation of the spectral data.

8.10. FIRAS Interpretation

When interpreting the implications of the FIRAS spectrum on effects that could distort the microwave background, it is important to remember some basic order-of-magnitude facts. Theoretical analyses of light element abundances gives limits on the current density of baryons through models of Big Bang NucleoSynthesis (BBNS). The ratio of baryons to photons[42] is $\eta = (3.5 \pm 0.7) \times 10^{-10}$, while another determination[43] gives $\eta = (3.95 \pm 0.25) \times 10^{-10}$. As a result, the effect of Lyman a photons from recombination is negligible, simply because there are so few hydrogen atoms per photon. Since the number density of photons is well determined by the FIRAS spectrum

$$N = 8\pi\zeta(3)\Gamma(3)\left(\frac{kT_0}{hc}\right)^3 = 410 \pm 5\,\text{cm}^{-3} \qquad (8.11)$$

the baryon density is well determined: 1.62×10^{-7} cm^{-3}. This gives $\Omega_B h^2 = 0.0144$. Not all of these baryons are protons, however. For a primordial helium abundance of 23.5% by mass, one has 13 protons and 1 *a* particle in 17 baryons. This gives an electron density of $n_{e,o} = 1.43 \times 10^{-7}$ cm^{-3}. For a fully ionized Universe, the optical depth for electron scattering over a path length equal to the Hubble radius c/H_0 is $0.00088/h$.

The expansion of the Universe H varies with time and redshift in a way that depends on ratios of the current matter, radiation and vacuum densities to the critical density $\rho_c = 3H^2/8\pi G$. Let these ratios be Ω_m, Ω_r and Ω_v. Then

$$H(z) = (1+z)H_0\sqrt{(1-\Omega_m - \Omega_r - \Omega_v) + \Omega_m(1+z) + \Omega_r(1+z)^2 + \Omega_v(1+z)^{-2}}$$

(8.12)

However, the expansion of the universe causes a significant change in the electron density in the light travel time. This effect must be taken into account.

For early epochs when the universe was ionized, Compton scattering is the dominant mechanism for transferring energy between the radiation field and the matter. As long as the electron temperature is less than about 10^8 K, the effect of Compton scattering on the spectrum can be calculated using the Kompaneets equation,

$$\frac{\partial n}{\partial y} = x^{-2}\frac{\partial}{\partial x}\left[x^4\left(n+n^2+\frac{\partial n}{\partial x}\right)\right]$$

(8.13)

where n is the number of photons per mode [$n = 1/(e^x - 1)$ for a black body], $x = h\nu/kT_e$, and the Kompaneets or Comptonization parameter y is defined by:

$$dy = \frac{dT_e}{m_e c^2}n_e\sigma_T cdt$$

(8.14)

Thus, y is the electron scattering optical depth times the electron temperature in units of the electron rest mass. Note that the electrons are assumed to follow a Maxwellian distribution, but that the photon spectrum is completely arbitrary.

The stationary solutions for the equation $\partial n / \partial T = 0$ are the photon distributions in thermal equilibrium with the electrons. Since photons are conserved, the photon number density does not have to agree with the photon number density in a black body at the electron temperature. Thus, the more general Bose–Einstein thermal distribution is allowed: n = 1/(exp(x + μ) - 1). This gives $\partial n / \partial T = 0$ for all μ. Since the Bose–Einstein spectrum is a stationary point of the Kompaneets equation, it is the expected form for distortions produced at epochs when

$$(1+z)\frac{\partial y}{\partial z} = \sigma_T n_{e,0} \frac{kT_0}{m_e c^2} \frac{c}{H} (1+z)^4 > 1 \qquad (8.15)$$

For $\Omega_B h^2$ given by BBNS, the redshift at the limits of this inequality is well within the radiation dominated era with a value $z_v = 10^{5.1}\sqrt{70\Omega_B h^2}$.

There is a simple solution to the Kompaneets equation with non-zero $\partial n / \partial T$ which gives the Sunyaev & Zeldovich (1972) or y spectral distortion. This simple case occurs when the initial photon field is a black body with a temperature 7_7 which is below the electron temperature. Letting $f = T_e/T_\gamma$, we find that the initial photon field is given by n = 1/(exp(fx) − 1). Therefore,

$$\left(n + n^2 + \frac{\partial n}{\partial x}\right) = \frac{(1-f)\exp(fx)}{(\exp(fx)-1)^2} = (1-f^{-1})\frac{\partial n}{\partial x} \qquad (8.16)$$

distortion has a Sunyaev–Zeldovich shape but is reduced in magnitude by a factor $(1 - T_\gamma/T_e)$. Defining the "distorting" y as

$$dy_D = \frac{k(T_e - T_\gamma)}{m_e c^2} n_e \sigma_T c dt \qquad (8.17)$$

we find that the final spectrum is given by a frequency-dependent temperature given by

$$T_v = T_0\left[1 + y_D\left(\frac{x(e^x + 1)}{e^x - 1} - 4\right) + ...\right] \qquad (8.18)$$

where $x = h\nu/kT_0$. The FIRAS spectrum in Fig. 8.2 shows that $|y_D| <$ 2.5×10^{-5}.

The energy density transferred from the hotter electrons to the cooler photons in the y distortion is easily computed. The energy density is given by $U \propto \int x^3 n dx$ so

$$\frac{\partial U}{\partial y_D} = \int x \frac{\partial}{\partial x}\left(x^4 \frac{\partial n}{\partial x}\right) dx \qquad (8.19)$$

Which when integrated by parts twice gives

$$\frac{\partial U}{\partial y_D} = -\int\left(x^4 \frac{\partial n}{\partial x}\right) dx = 4\int x^3 n dx = 4U \qquad (8.20)$$

Thus the limit on y_D gives a corresponding limit on energy transfer: $\Delta U/U < 10^{-4}$. Any energy which is transferred into the electrons at redshifts $z > 7$ where the Compton cooling time is less than the Hubble time will be transferred into the photon field and produce a y distortion. Since there are 10^9 times more photons that any other particles except for the neutrinos, the specific heat of the photon gas is overwhelmingly dominant, and the electrons rapidly cool (in a Compton cooling time) back into equilibrium with the photons. The energy gained by the photons is $\Delta U = 4yaT_\gamma^4$ which must be equal to the energy lost by the electrons and ions: $1.5(n_e + n_i)k\Delta T_e$. Since $y = \sigma_T n_e (kT_e/m_e c^2)c\Delta t$ we find the Compton cooling time

$$t_C = \frac{1.5(1 + n_i/n_e)m_e c^2}{4\sigma_T c U_{rad}} = \frac{7.4\times10^{19} \text{ sec}}{(1+z)^4} \qquad (8.21)$$

for an ion to electron ratio of 14/15. This becomes equal to the Hubble time $(3.08568 \times 10^{17}\text{sec})/(h(1 + z)^{1.5})$ at $(1 + z) = 9h^{0.4}$. Figure 8.3 shows the maximum allowed temperature for an intergalactic medium heated impulsively as a function of the heating redshift. After begin heated, the IGM cools by adiabatic expansion and Compton cooling. The 3 models shown in Fig. 8.3 are an open model with $H_0 = 100$ kms^{-1} Mpc^{-1}, $\Omega_B h^2 =$ 0.0125, and $\Omega = 0.2$; a flat model with $H_0 = 50$, $\Omega = 1$ and the same Ω_B; and a "BARYONIC" model with $H_0 = 50$ and $\Omega = \Omega_B = 0.2$ which has

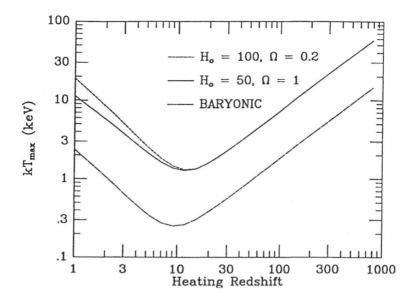

Fig. 8.3. The maximum allowed temperature for an IGM as a function of the heating redshift for two models with the BBNS baryon abundance ($\Omega_B h^2 = 0.125$) and one model with 4 times more baryons ($H_0 = 50$, $\Omega - \Omega_B = 0.2$).

four times more baryons than BBNS models predict. For this increased value of the baryon abundance, the allowed kT_{max} is considerably less than the 1 keV suggested by the 300 km/sec velocity dispersion in galaxies.

At redshifts $z > z_y = 10^5$, there will be enough electron scattering to force the photons into a thermal distribution with a μ distortion instead of a y distortion. However, the normal form for writing a μ, distortion does not preserve the photon number density. Thus we should combine the μ. distortion with a temperature change to give an effect that preserves photon number. The photon number density change with μ is given by

$$N \alpha \int \frac{x^2 dx}{\exp(x+\mu)-1}$$

(8.22)

with $x_0 = \sqrt{A}$. With this form for n one finds that

$$\frac{\partial N}{\partial y} = \int x^2 \frac{\partial n}{\partial y} dx = \int x^2 A \frac{1-e^{-x}}{x^3} \left(\frac{1}{e^x - 1} - n \right) dx$$

$$= A\mu / x_0 = \mu \sqrt{A}$$

(8.23)

in the limit that μ and x_0 are $<< 1$. In this limit the deficit of photons associated with μ is $\mu\pi^2/3$, we find a thermalization rate per unit y of

$$\frac{\partial \ln \mu}{\partial y} = \frac{3\sqrt{A}}{\pi^2} \alpha \sqrt{1+z}$$

(8.24)

Since $(1+z)\partial y / \partial z \propto \Omega_B h^2 (1+z)^2$, the overall rate for eliminating a μ distortion scales like $\Omega_B h^2 (1+z)^{5/2}$ per Hubble time. A proper consideration[44] of this interaction of the photon creation process with the Kompaneets equation and shows that the redshift from which $1/e$ of an initial distortion can survive is

$$z_{th} = \frac{4.24 \times 10^5}{\left[\Omega_B h^2 \right]^{0.4}}$$

(8.25)

which is $z_{th} = 2.3 \times 10^6$ for the BBNS value of $\Omega_B h^2$.

8.11. Summary

COBE has been a remarkably successful space experiment with dramatic observational consequences for cosmology, and the DIRBE determination of the cosmic infrared background is yet to come. The very tight limits on deviations of the spectrum from a blackbody rule out many non-gravitational models for structure formation, while the amplitude of the AT discovered by the *COBE* DMR implies a magnitude of gravitational forces in the Universe that is sufficient to produce the observed clustering of galaxies, *but only if the Universe is dominated by dark matter.*

Bibliography

1. Bennett, C. L. *et al.*, 1992, ApJL, 396, L7.
2. Bertschinger, E., Dekel, A., Faber, S. M., Dressier, A. & Burstein, D. 1990, ApJ, 364, 370–395.

3. Bennett, D.P. & Rhie, 1993, ApJ, 406, L7.

4. Bertschinger, E., Dekel, A., Faber, S.M. Dressier, A. & Burnstein, D. 1990, ApJ, 364, 370.

5. Boggess, N. *et al.*, 1992, ApJ, 397, 420.

6. Burigana, C, De Zotti, G. F., & Dáñese, L. 1991, *ApJ*, 379, 1–5.

7. Cheng, E.S., Cottingham, D.A., Fixsen, D.J., Inman, C.A., Kowitt, M.S., Meyer, S.S., Page, L.A., Puchalla, J.L. & Silverberg, R.F. 1993, ApJL, 422, L37.

8. Crittenden, R., Bond, J.R., Davis, R.L., Efstathiou, G. & Steinhardt, P.J. 1993, Phys, Rev. Lett., 71, 324.

9. de Bernardis, P., Masi, S. Melchiorri, G., Melchiorri, B. & Vittorio, N. 1992, ApJ, 396, L57.

10. Devlin, *et al.*, 1994, submitted to ApJ.

11. Dragovan, M. *et al.*, 1993, preprint.

12. Fich, M., Blitz, L. & Stark, A.A. 1989, ApJ, 342, 272.

13. Fixsen, D.J. *et al.*, 1994a, ApJ, 420, 445.

14. Fixsen, D.J. *et al.*, 1994b, ApJ, 420, 457.

15. Gaier, T. *et al.*, 1992, ApJ, 398, LI.

16. Ganga, K. *et al.*, 1993, ApJ, 410, L57.

17. Gorski, K. 1992, ApJ, 398, L5.

18. Gundersen, J., Clapp, A., Devlin, M., Holmes, W., Fischer, M., Meinhold, P., Lange, A., Lubin, P., Richards, P. & Smoot, G. 1993, ApJ, 413, LI.

19. Gush, H.P., Halpern, M. & Wishnow, E.H. 1990, Phys. Rev. Lett., 65, 537.

20. Guth, A. & Pi, Y.-S. 1982, Phys, Rev. Lett., 49, 1110.

21. Hancock, S. *et al.*, 1994, Nature, 367, 333.

22. Hawking, S. 1982, Phys. Lett., 115B, 295.

23. Johnson, D. & Wilkinson, D.T. 1989.

24. Kerr, F.J. & Lynden-Bell, D. 1986, MNRAS, 221, 1023.

25. Kneissl, & Smoot, G.F. 1993, COBE memo/Santarder Conf. proceedings.

26. Kogut, A. *et al.*, 1992, ApJ, 401, 1.

27. Kogut, A. *et al.*, 1993, ApJ, 419, 1.

28. Lauer, T. & Postman, M. 1992, BAAS, 24, 1264.

29. Martin, D.H. & Puplett, E. 1970, Infrared Physics, 10, 105.

30. Mather, J.C. *et al* 1990, ApJ, 354, L37.

31. Mather, J.C. *et al.*, 1994, ApJ, 420, 439.

32. Meinhold, P., Clapp, A., Devlin, M., Fischer, M., Gundersen, J., Holmes, W., Lange, A., Lubin, P., Richards, P. & Smoot, G. 1993, ApJ, 409, LI.

33. Myers, ST., Readhead, A.C.S. & Lawrence, C.R. 1993, ApJ, 405, 8.

34. Partridge, R.B. & Peebles, P.J.E. 1967, ApJ, 148, 377.

35. Peterson, J. *et al.*, 1994, Astrophysics Letters and Communication.

36. Readhead, A.C.S., Lawrence, C.R. Myers, S.T., Sargent, W.L.W., Hardebeck, H.E. & Moffet, A.T. 1989, ApJ, 346, 566.

37. Sachs, R.K. & Wolfe, A.M. 1967, ApJ, 147, 73.

38. Schuster, J., Gaier, T., Gundersen, J., Meinhold, P., Koch, T., Seiffert, M., Wuensche, C. & Lubin, P. 1993, ApJ, 412, L47.
39. Smoot, G.F. & Steinhardt, P. 1993, J. Quantum & Classical Gravity, 10, 1993.
40. Smoot, G.F. *et al.*, 1990, ApJ, 360, 685.
41. Smoot, G.F. *et al.*, 1991, ApJ, 371, LI.
42. Staggs, S. 1994, Astrophysics Letters and Communication, to be published.
43. Starobinskii, A.A. 1982, Phys. Lett., 117B, 175.
44. Stecker, F.W., Puget, J.L. & Fazio, G.G. 1977, ApJ, 214, L51.
45. Subrahmayan, R., Ekers, R.D., Sinclair, M. & Silk, J. 1993, MNRAS, 263, 416.
46. Sunyaev, & Zeldovich, 1972, Cos. Ap. Space Science, 4, 173.
47. Walker, T.R, Steigman, G., Schramm, D.N., Olive, K.A. & Kang, H.-S. 1991, ApJ, 376, 51.
48. Watson, R.A., Gutierrez de la Cruz, CM., Davies, R.D., Lasenby, A.N., Rebolo, R., Beckman, J.E. & Hancock, S. 1992, Nature, 357, 660.
49. Wollack, E.J., Jarosik, N.C., Netterfield, C.B., Page, L.A. & Wilkinson, D.T. 1994, ApJ, in press.
50. Wright, E.L. 1993 in The 16th Texas Symposium, Annals of the New York Academy of Sciences, in press.
51. Wright, E.L. *et al.*, 1992, ApJ, 396, L13.
52. Wright, E.L. *et al.*, 1994a, ApJ, 420, 1.
53. Wright, E.L. *et al.*, 1994b, ApJ, 420, 450.
54. Yahil, A., Tammann, G.A. & Sandage, A. 1977, ApJ, 217, 903.

The Last Thirteen Billion Years…
(Krakow, 1998/Madrid, 2002)

Chapter 9

Unexpected Coincidence between Decoupling and Atom Formation Times

ACTA COSMOLOGICA   FASCICULUS XXIV–2 1998

UNIVERSITAS IAGELLONICA—ACTA SCIENTIARUM LITTERARUMQUE MCCXXVIII

HOW CLOSE ARE THE COSMIC TIMES FOR MATTER/RADIATION EQUALITY AND FOR ATOM FORMATION?

Noé Cereceda, Ginés Lifante and Julio A. Gonzalo

Facultad de Ciencias, C-IV, Universidad Autónoma de Madrid
28049 Madrid, Spain

(Received September 24, 1997; final version March 15, 1998)

Abstract. It is shown that the cosmic matter/radiation equality time ($\rho_r = \rho_m$) corresponding to $H_o t_o \cong 0.910$, and the atom formation time as determined from Saha's law, appear to coincide almost exactly at a background radiation temperature about $T_{eq} \cong T_{af} \cong 3840$ K. The resulting maximum amount of dark matter comes out to be of the order of that of ordinary baryonic matter which is consistent approximately with the lower limit of Ω_o estimated from gravitational lenses. We get $\Omega_o \cong 0.0824$ (total) to be compared with $\Omega_{bo} \cong 0.0355$ (baryonic).

In 1998 the following paper by N. Cereceda, G. Lifante and J. A. Gonzalo, was published in "Acta Cosmologica". At that time observational estimates of Hubble's parameter (H₀) ranged between 50 and 100 km/sMpc, and current estimates of the time elapsed since the Big-Bang (to) between 20 and 10 billion years. Using properly the Friedmann-Lemaitre open solutions for Einstein cosmological equations we got H_0 = 65 km/sMpc and t_0 = 13.7 × 10^9 years. Five years later, in February 2003, the New York Times reported WMAP's finding: H_0 = 69 ± 2 km/sMpc, t_0 = 13.7 ± 0.2 billion years.

9.1. Introduction

It is well known that radiation/matter equality ($\rho_r = \rho_m$) and atom formation (ions + electrons → atoms) occur at cosmic times (temperatures) relatively close to each other. How close are the radiation/matter equality (t_{eq}) and the atom formation (t_{af}) times? Is there an approximate coincidence between them or a nearly exact one? Is it fortuitous or is there a direct physical connection between both? Current estimates[1] give $t_{af} \sim 3 \times 10^5$ yrs, not far, but substantially larger, than $t_{eq} \sim 1 \times 10^4$ yrs. In this work, we argue that, in spite of the present uncertainties in our knowledge of H_0 (present value of the Hubble's constant) and t_0 (time elapsed since the Big Bang) their product is related by means of Einstein's equations[2] to Ω_0 (present value of the cosmic density parameter). This constrains the ranges of acceptable values for H_0, to and Ω_0. From these considerations alone we show that the resulting time for matter/radiation equality is in almost perfect coincidence the atom formation time, given by $t_{eq} \approx t_{af} \approx 3 \times 10^5$ yrs. The corresponding contemporary CBR temperature is T ≈ 3840 K. After atom formation the universe is made up of atoms (transparent to radiation), and prior to it is made up of a plasma; of ions and electrons (opaque to radiation). In the present work, we put aside the dynamics of cosmic evolution prior to matter density/radiation density equality (t_{eq}) leaving it for future consideration elsewhere).

The solutions of Einstein equations,[2] relate parametrically the time (t) and the scale factor (R), which is in turn related to the temperature (T) by the appropriate equation of state. In the present, matter dominated era ($\rho_r = \rho_m$) of the cosmic expansion, the cosmic equation of state is given by RT≈const. Prior to atom formation the equation of state is generally assumed to be the same, but, as mentioned above, we will not be concerned here with this assumption.

We set the open solutions[3,4] of Einstein cosmological equations in a convenient form to get (Ht) and (Ω) a function of $R/R_+ = T_+/T$, using the equation of state for a transparent universe, where R_+, and T_+ are taken as the scale factor and temperature, respectively, at which the terms involving the total mass density and the space curvature in Einstein equation have the same value, i.e. at $(8\pi G/3)pR_+^3 = c^2|k|R_+$. Then we use observational constraints on H_0, t_0 and T_0 (CBR temperature) to fix T_+, knowing very well, as we do, the present value of the CBR temperature T_0.[5] From this starting point we will get the value of the radiation/matter equality temperature, T_{eq}, at which the mass densities of radiation (ρ_r) and matter (ρ_m) become equal. Next we evaluate the atom formation temperature, T_{af}, from Saha's law, under the assumption that the matter mass density ρ_m (related to the particle density $n_p \approx \rho_m/m_p^*$, where m_p^* is the effective ion mass) is equal to the radiation mass density, given by Stefan–Boltzmann law as $\rho_0 = \sigma T^4/c^2$. We show that the values for T_{eq} and $T_x = T_{af}$ thus obtained are in almost perfect coincidence with each other.

9.2. Radiation/Matter Equality Temperature

In the time interval between t_0 (present) and t_x (atom formation), near to or equal to decoupling, the equation relating the change in internal energy of the cosmos with the work done in the expansion leads to

$$\rho = \rho_m + \rho_r = \rho_m\left(1 + \frac{R_x}{R}\right) \tag{9.1}$$

where

$$\rho_r R^4 = \left(\frac{\sigma T^4}{c^2}\right)R^4 = \text{const} \tag{9.2}$$

and

$$\rho_m R^3 = \text{const} \tag{9.3}$$

implying $RT = \text{const}$ and $\rho_x = 2\rho_{mx} = 2\rho_{ra}$ (at t_{eq}.).

Einstein's field equation[3,4] for an open (k < 0) or flat (k = 0) universe is given by

$$\dot{R} = R^{-1/2}\left(\frac{8\pi G}{3}\rho R^3 + c|k|R\right)^{1/2} \qquad (9.4)$$

Using Eq (9.1) for $t > t_x$ we get

$$\int dt = \int\left[\left(\frac{8\pi G}{3}\rho_{m+}R_+{}^3\right)(R_+ + R) + c^{22}|k|^2 R^2\right]^{-1/2} R\,dR \qquad (9.5)$$

where, as stated before, R_+ is fixed at $(8\pi G/3)\rho_+ R_+{}^3 = c^2|k|R_+$, and $R = R_o(T_o/T)$ will be the scale factor at any time t in the interval $t_{eq} < t <$ to.

Integration of Eq. (9.5) gives

$R = R_+ \sinh^2 y$

$$t - t_x = \frac{R_+}{c\sqrt{|k|}}\left[\left\{\frac{R_x}{R_+} + \sinh^2 y \cosh^2 y\right\}^{1/2} - \left\{\frac{R_x}{R_+} + \sinh^2 y_x \cosh^2 y_x\right\}^{1/2}\right.$$

$$\left. -\frac{1}{2}\ln\left[\frac{1}{2} + \sinh^2 y + \left\{\frac{R_x}{R_+} + \sinh^2 y \cosh^2 y\right\}^{1/2}\right]\right/$$

$$\left[\frac{1}{2} + \sinh^2 y_x + \left\{\frac{R_x}{R_+} + \sinh^2 y_x \cosh^2 y_x\right\}^{1/2}\right]\right]$$

For $R_x/R_+ = T_+/T_x \ll 1$ the above time equation becomes,

$$t - t_x = \frac{R_+}{c\sqrt{|k|}}\left[\sinh y \cosh y - y\right] \qquad (9.6)$$

Using $R(y)$ and $t(y)$ expressions for the Hubble parameter H, and the dimensionless cosmic parameters, (Ht) and Ω, are obtained, as

$$H = \frac{\dot{R}}{R} \cong \frac{c\sqrt{|k|}}{R_+}\frac{\cosh y}{\sinh^3 y} \qquad (9.7)$$

$$Ht = \frac{\dot{R}}{R}t \cong \frac{1}{\tanh^2 y} - \frac{y}{\tanh y \sinh^2 y} \qquad (9.8)$$

$$\Omega = \frac{\rho}{\dfrac{3H^2}{8\pi G}} \cong \frac{1}{\cosh^2 y} \tag{9.9}$$

Further, if at t_x (radiation/matter equality) the temperature T_x is $T_x \gg T_+$

$$y_x = \sinh^{-1}\left(\frac{R_x}{R_+}\right)^{1/2} = \sinh^{-1}\left(\frac{T_+}{T_x}\right)^{1/2} \ll 1 \tag{9.10}$$

Eq. (9.6) for tx becomes, to a very good approximation,

$$t_x \cong \frac{R_+}{c\sqrt{|k|}} \frac{2}{3} y^3 \tag{9.11}$$

which can be put into the alternative form

$$t_x \cong \frac{2}{3} H_x^{-1} \Omega_x^{-1/2} = \frac{2}{3} H_0^{-1} \Omega_0^{-1/2} \left(\frac{R_x}{R_0}\right)^{3/2} = \frac{2}{3} H_0^{-1} \Omega_0^{-1/2} \left(\frac{T_0}{T_x}\right)^{3/2} \tag{9.12}$$

where $(R_x/R_0) = (T_0/T_x) = (1 + z_x)^{-1}$, being z_x the redshift at time t_x.

Basic known cosmic parameters at present times are the CBR temperature $T_0 = 2.726 \pm 0.01$ K, the lower limit to the Hubble constant $H_0 > 65$ km/s.Mpc from recent Hubble telescope data on classical Cepheid variables,[6] and a lower bound to $t_0 > 13.7 \times 10^9$ yrs[7] based upon a re-evaluation of the age of the oldest globular clusters (main sequence turn off). The above values of H_0 and t_0 imply.

$$H_0 t_0 \geq 65(\text{km/s.Mpc}) \times 13.7 \times 10^9 \text{yrs} = 0.910 \text{ (dimensionless)} \tag{9.13}$$

Equations (9.8) and (9.9) on the other hand connect ($H_0 t_0$) and Ω_0 through the parameter $y_0 = \sinh^{-1}(R_0/R_+)^{1/2}$ restricting $H_0 t_0$ to be in the interval $(2/3) < H_0 t_0 < 1$, with $(2/3)$ for a flat universe ($k = 0$, $\Omega_0 = 1$) and $H_0 t_0 \rightarrow 1$ for an open universe ($k < 0$, $\Omega_0 < 1$) with a vanishing amount of matter. We may note that a very recent report, (Phys. Today, Sept. 1997, pp. 19–21) of the careful analysis of data compiled on stars out to an unprecedented distance of about 500 light-years, taken by the European Space Agency Hipparcos Satellite, basically confirms that to must be slightly larger than 13×10^9 years. This analysis appears to reconcile the estimated values of to from globular stars clusters in our galaxy and from the Hubble's constant.

Figure 9.1 gives Ω_0 vs. $H_0 t_0$ showing $(H_0 t_0)_{min} \cong 0.910$ and the estimated present baryonic mass density parameter $\Omega_{b0} = 0.0355 <$ $(\Omega)_{max} \cong 0.0824$, taking $(\Omega_0)_{max}$ as fixed bu Eq. (9.13). Since observational estimates on the total matter mass density parameter[8] from gravitational lenses give $0.1 < \Omega_0 < 0.3$, our upper bound to $(\Omega)_{max} \approx 0.084$ comes out to be close to $\Omega_0 = 0.1$ and we will use it in our analysis of cosmic evolution from radiation/matter equality/atom formation, t_x, to present times.

From the above value for $(H_0 t_0) \approx 0.910$, using RT = const as the equation of state for a transparent universe $(t \geq t_x)$, we get

$$y_0 = \sinh^{-1} \left(\frac{T_+}{T_x} \right)^{1/2} \geq 1.92, \quad \text{i.e. } T_+ \geq 30.4K \qquad (9.14)$$

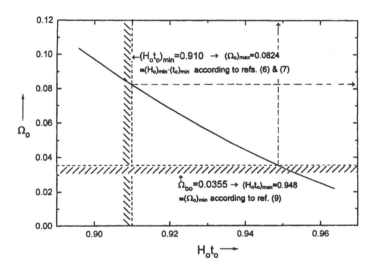

Fig. 9.1. $\Omega_0 \cong 1/\cosh^2 y$ vs. $H_0 t_0 \cong (\sinh y_0 \cosh y_0 - y_0)\cosh y_0 / \sinh^3 y_0$. Vertical dashed line gives constrain on the baryonic mass density $\Omega_{b0} \approx 0355$, from [4]He abundance fixed from the time of nucleosynthesis.

Once we have a numerical value $T_+ = 34.2$ K we are in a position to obtain numerical values for $t(T)$, $\rho_r(T)$, $\rho_m(T)$ in terms of $y(T)$ at any $T < T_x$.

At radiation/matter equality, $\rho_{req} = \rho_{meq}$, which implies

$$\frac{\sigma T_{eq}^4}{c^2} = \rho_{m0}\left(\frac{T_{eq}}{T_0}\right)^3 \tag{9.15}$$

where $\rho_{m0} \approx \Omega_0 3 H_0^2 / 8\pi G = 6.54 \times 10^{-31}$ g/cm³, $\Omega_0 = 0.0824$ and $H_0 = 2.10 \times 10^{-18}$, one gets

$$T_{eq} = \frac{\rho_{m0}}{\dfrac{\sigma T_0^3}{c^2}} - 3.84 \times 10^3 \, K \tag{9.16}$$

This value becomes close to the temperature near the Sun's surface, where ionisation and recombination of electrons and positive hydrogen and helium ions are known to be taking place.

9.3. Atom Formation

Now we can try to exploit Saha's law for ionization/recombination in a plasma to get independent estimate of T_x under the assumption that the radiation and the matter energy densities in the plasma are equal. Taking into account that the relative abundance of H and ^{4}He (ignoring other very scarce light elements) is well established, we can estimate the effective average mass per ion as $m^* = m_{eff} = 1.187 \, m_p$ and the effective average ionization energy as $B_{eff} = 14.27$ eV. This is done by noting[9] that the cosmic neutron to proton, ratio (n/p) is determined by

$$(n/p) \cong \frac{1}{2}\frac{Y_p}{1 - \dfrac{1}{2}Y_p} \cong 0.131, \quad Y_p = 0.232 \pm 0.008 \tag{9.17}$$

Implying that the corresponding average fractions of protons and ^{4}He nuclei are given by

$$f_H \cong 0.9374, \quad f_{^4He} \cong 0.0626, \tag{9.18}$$

and that average ionization energy is, therefore,

$$B_{eff} = f_H(13.6eV) + f_{^4He}(22.4eV) = 14.27eV. \tag{9.19}$$

Saha's Eq. (9.10) can be written as

$$\frac{x^2}{1-x} = \frac{1}{n_{eff}} \left(\frac{m_e k_B T}{2\pi \hbar^2} \right)^{3/2} e^{-B_{eff}/k_B T} \qquad (9.20)$$

where x is the fraction of ionised atoms, equal to 50% right at T_x (the ionisation/recombination temperature), n_{eff} is the effective average ion density, given for $\rho_{mx} = \rho_{rx}$ by

$$n_{eff} = \frac{\Omega_{b0}}{\Omega_0} \frac{\rho_{mx}}{1.187 m_p} = \frac{\Omega_{b0}}{\Omega_0} \frac{\dfrac{\rho_{mx} \sigma T_x^4}{c^2}}{1.187 m_p} \qquad (9.21)$$

whit $(\Omega_{b0}/\Omega_0) = 0.0355/0.0824$, and B_{eff} is the effective average ion density, given by Eq. (9.19). Rearranging terms we get

$$3.716 \times 10^{-15} = \left(\frac{B_{eff}}{k_B T_x} \right)^{5/2} e^{-B_{eff}/k_B T} = \beta_x^{5/2} e^{-\beta_x} \qquad (9.22)$$

This results in

$$\beta_x = \frac{B_{eff}}{k_B T_x} = 42.88,$$

and therefore, using $B_{eff} = 14.27 eV$,

$$T_x = 3880 \text{ K}, \qquad (9.23)$$

which is in excellent agreement with $T_{eq} = 3840$ K obtained above from purely cosmological quantities. The very good agreement is a little surprising, considering the inevitable approximations and guesswork involved in determining T_{eq}. It must be emphasized that the method used is simple and direct, and that use of the best available observational evidence, as far as we know, has been made. In Fig. 9.2 we have plotted x as a function of T according to Eq. (9.20) for $B_{eff} = 14.27$ eV (H, ^{4}He) and for B = 13.6 eV (H). The difference is small but significant. Note also that the transition between 90% ionisation and 10%) ionisation occurs sharply within. ±8.7% of T_X = 3880 K, disregarding other suggested possible complications.[4]

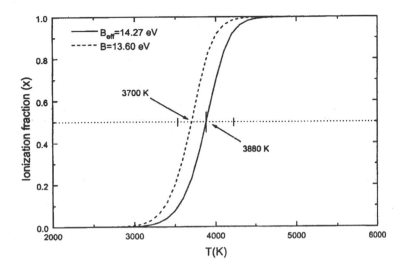

Fig. 9.2. x (ionisation fraction) vs. T (temperature) under the assumption $\rho_{mx} = \rho_{rx}$, for $B_{eff} = 14.27$ eV (H, ^{4}He) and for B = 13.6 eV, according to Sana's equation.

9.4. Concluding Remarks

The coincidence found between radiation/matter equality temperature and atom formation temperature raises the question of the proper equation of state prior to the atom formation time. We will take up this point in future work elsewhere.

Table 9.1 gives the dimensionless cosmic parameters Ω = matter mass to critical mass ratio, (Ht) = Hubble's parameter times cosmic time, and (ρ_r/ρ_m) = radiation to matter densities ratio, all as function of the contemporary CBR radiation temperature. Note that in the inflationary paradigm $\Omega = 1$, Ht = 2/3, at all T.

Table 9.1. Cosmic parameters as a function of CBR radiation temperature.

Temperature (K)	$T_0 = 2.726$	$T_+ = 30.4$	$T_{eq} = 30.4$	$(T \gg T_{eq})$
Ω	0.0824	0.500	0.992	≈ 1
(Ht)	0.910	0.733	0.667	$\approx 2/3$
(ρ/ρ_m)	7.1×10^{-4}	7.31×10^3	1	–

We conclude that a critical evaluation of the available observational evidence supports a close coincidence between the cosmic matter/ radiation equality time and the atom formation time.

Bibliography

1. BENNET, C. L., TURNER, M. S. and WHITE, M. *Physics Today*, Nov. (1997) Box 1: Big Bang Basics, p. 34.
2. FRIEDMAN, A. Z., *Physik* 10, 377 (1992); LEMAITRE, G., "An. Soc. Sci. Bruxelles", 47A, 49 (1927).
3. WEINBERG, S., *Gravitational Cosmology* (N.Y. Wiley and Sons, 1972).
4. PEEBLES, P. J. E., *Principles of Astrophysical Cosmology* (Princeton University Press, 1993).
5. MATHER, J. C. *et al.*, *Astrophys. J.* 429, 439–444 (1994).
6. FREEDMAN, W. L. *et al.*, *Nature*, 371, 757–762 (1994).
7. BOTTE, M. and HOGAN, C. J., *Nature*, 376, 399–402 (1995).
8. SCHRAMM, D. N., *Physica Scripta*, Vol. T36, 22–29 (1991).
9. OLIVE, K. A. and STEIGMAN, *Astrophysics.*, *J. Suppl. Ser.* 97, 49 (1995).
10. DENDY, R. (Ed), *Plasma Physics: An introductory Course* (Cambridge University Press, 1993).

Chapter 10

An Amazing Story: From the Cave Man to the Apollo Mission

From the most remote times men have felt the fascination of the starry night and have noted with astonishment the regularities which characterize the motions of the Sun, the Moon and the Planets through heaven. We are entitled to say that astronomy is the most ancient pure science, and, like many other pure sciences, it found soon practical applications, as making calendars, useful to organize agricultural work, to register the chronicles of past times, and to do plans form the future.

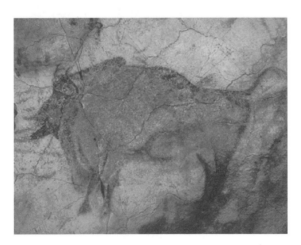

Fig. 10.1. Altamira Cave.

From the dawn of historical times[1] we have received calendars made by astronomers-astrologers (at that time there was not a clear cut distinction between both) from Babylon, Egypt and China, as well as from the Americas, like the ones elaborated by the Mayas and the Aztecs.

Contemporary astrophysical cosmology which, for the first time in history, has attempted a systematic and coherent description of the totality of matter and energy in the Universe and their interactions has been successful in achieving an harmonic combination of precise observations and elegant mathematical theory to make up a unitary scientific discipline. As professor Murray Gell–Mann, winner of the Physics Nobel Prize in 1969 said[2] some time ago it is surprising that a handful of beings, inhabitants of a small planet circling an insignificant star, have been able to trace back their origin to the beginning, that a "grain of dust" has been able to grips "the entire universe".

Fig. 10.2. Aztec Calendar.

The amazing intellectual adventure begins with Erathostenes an Hipparcos. Goes from Buridan to Copernicus and Kepler. Receives decisive impetus with Newton. Reaches maturity with Einstein and Lemaitre, on one side, and with Hubble, Penzias and Wilson on the other. Torwards the end of the 20th century, through large scale space projects such as NASA's "COBE" (Cosmic Background Explorer), the "Hubble" space telescope, and the European space mission "Hipparcos", it culminates in a coherent picture of the physical universe. All of it, together with to day's current projects such as the one systematically observing Type Ia Supernovae, allow us to follow in broad strokes the thermal history of the universe from the Big Bang to the present epoque. A universe formed by material particles (grouped in galaxies stars and cosmic dust) and by pure radiation (photons) everything in rapid radial expansion in a grand cosmic scale.

Fig. 10.3. A. Einstein.

Einstein's cosmological equations, supplemented by reasonable estimates for the time elapsed since the Big Bang ("age" of the universe) and for the Hubble parameter ($H_0 = \dot{R}_0 / R_0$) giving the present ratio of the local expansion speed to the cosmic radius (measured from the center of expansion of the cosmic background radiation) allow one to detect a sequence of events, from the beginning (Big Bang) to the present epoque, events which have taken place as the cosmos was expanding and getting cooler. In the first few minutes, as beautifully described in a famous book, Steve Weinberg[3] (another distinguished Physic Nobel Prize winner), the primordial nucleosynthesis of ^{4}He and other light elements took place from protons and neutrons in thermal equilibrium at temperatures of the order of $T_{ns} \sim 10^8$ K.

Approximately half a million years after the Big Bang, the cosmic plasma made up of protons ⁴He nuclei and electrons was cool enough to give rise to bound atoms. The universe then became transparent ⁴He nuclei and electrons were cool enough to give rise to bound atoms. When the universe became transparent,[4] and under the attraction of gravity, protostars and protogalaxies begun to form.

From the time of the first galaxies and stars to the present époque, approximately thirteen billion (13×10^9) years have elapsed. During this long time span the universe has not ceased to expand and cool.

All along, our galaxy, the Via Lactea or Milky Way, acquired its present shape. The Sun, as a second or third generation star, formed out of cosmic dust and acquired its present form. The solar system, formed by the planets (in greek "planet" = vagabond), including the Earth-Moon system, our "habitant", took form.

About one billion years after the Earth took form, the physical conditions (hydrosphere, atmosphere) required to house living organisms (perhaps blue and green algae) were in place. And life begun on Earth. An event no less astonishing than the Creation of the Universe.

Fig. 10.4. Andromeda.

The observable universe is finite, i.e. it has an enormous mass, but finite a very large spatial extent, but finite, and, due to the fact that even if the expansion speed has been very large, being finite, the time elapsed from the beginning till now must have been necessarily finite. [Of course, we cannot observe infinite quantities of mass, length or time, even with the wonderful instruments presently at our disposal].

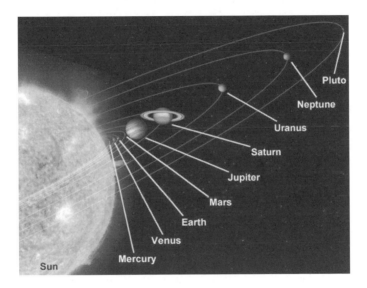

Fig. 10.5. Solar System.

If we estimate the number of observable galaxies in 10^{11} to 10^{12}, and the number of stars in each one of then in 10^{11} to 10^{12}, with an average mass of the order of the solar mass (2×10^{33} g) the resulting total number of massive particles in the universe, with a mass $m_p = m_n \approx 1.67 \times 10^{-23}$ g (the proton/neutron mass), is of the order of 10^{78} to 10^{80} particles. Enormous, evidently, but finite.

We cannot discard that more material particles of some kind are filling the universe. For the moment these particles have not been found. In the mean time a frantic search goes on, but even if great quantities of "dark matter" exists in the universe, it has been estimated in ten times, at most thirty times, the estimated baryonic mass actually existing as protons and heavier nuclei. And multiplying a finite number by ten or thirty would still give a finite number.

Our Via Lactea[5] is one of those great disk shaped spiral galaxies so abundant in the universe. It is very similar to the Andromeda galaxy, our close companion. With it, and other thirty smaller nearby galactic systems, forms what is known as the Local Group.

The diameter for the Via Lactea is about 30 kpc (1pc = 3.26 light years = 3.08×10^{18}m).

The approximate distance from the Sun to the center of gravity of the Via Lactea is about 8 to 9 kpc. Our galaxy, like other analogous spiral galaxies is rotating around its center of mass at about 220 km/sec. in the point corresponding to the Sun. A full rotation, i.e. a rotation of 2π radians, takes place in about 250×10^6 years. This means that, since its formation, about 13×10^9 years ago it has gone over more than 50 full rotations. Its central mass, leaving aside its massive halo, is about 10^{11} solar masses. The proportion of interstellar to stellar mass within the galaxy is about 10%. The data supplied by the space mission "Hipparcos" provide a priceless information on our Via Lactea, much more precise than available until recently.

The most important star for us is, of course, the Sun. Many national or supranational flags, such as those of the United States of America or the European Union use symbolically stars for their constituent units. My good friend Professor Kenkichi Okada, from the Nagoya Institute of Technology, used to say that for him the most beautiful of all these flags was the flag with a single star, the Rising Sun.

Our planet, the "blue planet", together with the other planets, circles the Sun. The names of Buridan, Oresme, Corpenicus, Galileo, Tycho Brahe and Kepler signal the way to the definitive description of the planetary motions, the heliocentric theory which would allow Newton to formulate his famous law of universal gravitation.

And Newton said[6] in his "Principia": "This most beautiful system of Sun, Planets and Comets could only proceed from the counsel and dominion of an intelligent and powerful Being".

Some philosophers of the Enlightenment, including Immanuel Kant, took for granted that any and all planets of our solar system were inhabited by beings with different degrees of intelligence. According to

Kant, the inhabitants of Venus, with a torrid climate (because of its closer proximity to the Sun) should be brutes compared with men, while the inhabitants of Mars (further from the Sun and cooler than the Earth) must be geniuses, compared with whom Newton's could only be considered a crude and vulgar intelligence.[7]

In the latest decades of the 20th century NASA's space probes have been unable to detect life at these planets, to say nothing of intelligent life.

A list can be made with the main cosmic pre-conditions necessary to support life, the type of life which is abundantly evident in our planet today:

1st) The numerical values of the physical universal constants quantitatively determining the four fundamental interactions (gravitational, electromagnetic, nuclear strong, and nuclear weak).

2nd) The primordial nucleosynthesis of ^4He in the adequate proportion at temperatures of the order of 10^8 K and determinant of the nuclear reactions within the first protostars.

3rd) The actual existence of massive stable nuclear isotopes synthesized in the center of stars.

4th) The actual existence of stable atoms at temperatures below 4×10^3 K.

5th) The continued existence of stars burning fist hydrogen for billions of years, then helium and so forth.

6th) The previous existence of very massive protostars capable of producing relatively light elements such as C, O, N, S, necessary to form part of planets as constituents of organic living matter.

7th) The simultaneous existence of heavier elements like Fe, Ni, Co capable of forming a central magnetic nucleus of an Earth-like planet, and others like Mg, Ca, Sr, Ba...capable of forming the external mantle.

8th) The formation of a planetary body with sufficient mass at an adequate distance form the central star (not too far, not too close) so that the mean temperature in the planet surface is of the order of 300° K, intermediate between the melting and boiling points of water H_2O.

9th) The requirement that the planet has a <u>minimum mass</u> to hold an atmosphere rich in O_2 and N_2.

10th) That the planet in question (the Earth) has a large satellite capable of stabilizing its rotation around an axis at an angle adequate to insure the seasonal cycles, avoiding extreme maxima and minima for the planet surface temperature.

11th) That it has a large planetary companion, such as Jupiter, capable of protecting it from large asteroids and meteors.

12th) The presence of ions in the high atmosphere (ionosphere) capable, under the action of the planetary magnetic filed, of making up some kind of van Allen belt protecting the planet from too much cosmic radiation. If the Sun, instead of being located in the outskirts, were close to the galaxy center, cosmic radiation would kill all life in the planet.

13th) The existence in the planet of abundant water whose molecule has such exceptional chemical properties that no other substance has even remotely regarding fitness to the elemental activities of living systems.

14th) The possible existence (and this is truly astonishing) of actual self-replicating macromolecules, such as DNA, with a characteristic structure and a sequence of components, A = Adenine, G = Guanine, C = Citosine, T = Timine, capable of making possible the complex diversity of life as seen on Earth today.

The list could go on and on, with no end in sight.

In particular, without the actual possibility of photosynthesis, which synthesizes organic matter taking advantage of the continuous flux of solar energy, consuming CO_2 and liberating O_2, vegetation and animal life on the planet depending on it would be impossible.

The simplest organisms are unicellular, but even the most elementary cell is a tremendously complex organism. And life itself is very difficult to define: a cell just before dying is almost chemically undistinguishable form a cell just after dying.

Around the middle of the 20th century, X-ray diffractometry, an experimental technique which von Laue, the Bragg's (father and son) and Debye pioneered, began to be applied to study the structure of organic matter. Basic discoveries for Molecular Biology followed in quick succession: the double helix structure of DNA (Watson & Crick,

1953), the structure of myoglobine, the first protein (Sanger, 1957) and the structure of Hemoglobine (Perutz, 1959).

All living organisms, from the most elementary bacteria to the most complex and developed mammal, are organisms that possess their characteristic DNA, with its genetic code, capable of ordering its development, with the corresponding mRNA and proteins, which control its metabolism, its immunologic defenses and its reproductive system. All are made up of the same basic elementary components. The mRNA, carrier of the information contained in the DNA from the nucleus to other locations in the cell, has a structure analogous to that of the DNA, and the proteins are macromolecules made up of a central nucleus of hydrophobic aminoacids and a long chain of hydrophilic aminoacids folded irregularly, with successive segments of the long chain linked by hydrogen bonds at specific far away points. There are a total of twenty aminoacids, each with an amino group (NH_2) and a carboxyl group (COOH) connected by a CH group from which hangs a chain, different for each aminoacid, which gives it its specific biochemical properties.

When did life appear on the Earth? We know that the initial atmospheric and its physical conditions were not hospitable to life when it was formed, 4.6×10^9 years ago. However, there are fossil remnants which indicate that 3.5×10^9 years ago there was life already.

Francis Crick, famous biochemist and Nobel Prize winner, in this book "Life Itself", says: "An honest man, armed with all the knowledge available to us now, could only state that in some sense, the origin of life appears at the moment to be almost a miracle, so many ate the conditions which would have had to have been satisfied to get it going".

This does not exclude the possibility that some future discovery might clarify the conditions under which life could appear on Earth, and even the conditions under which and evolution process leading to the higher vertebrates could be envisaged. But, in all appearance, this process could not be a blind, random process. Everything points to a contingent, therefore not necessary process, made possible by some extraordinarily favorable initial conditions.

Freeman Dyson, theoretical physicist and Templeton Prize winner, nothing that many chemists and biochemists believe that there may be millions of planets in our galaxy with intelligent habitants which might have achieved advanced technologies like ours, concludes, after careful consideration, that noting resembling a massive technological conquest has taken place in our galaxy, making unlikely the existence of such intelligent begins.

Paul Dirac (1902–1984), one of the pioneers of Quantum Mechanics, which in his youth is on record as leaning to atheist and Marxism, said with reference to the privileged and exceptional character of the Earth as a habitable planet, at a congress help in Lindau (1972) that if the occurrence of other inhabited planets in the cosmos were a frequent occurrence, it would speak against the existence of God, while, if it were something exceptional, unique, it would provide a powerful argument to affirm God's existence (Quoted by A. Herfiel, in "La Ciencia actual y Dios", Madrid: 1977).

The first fossil remnants attributed to unicellular living beings have been found in Australian rocks of some three and a half billion years of age containing layers of carbonaceous material. They correspond to organisms with a biochemical complexity comparable to that of living organisms now.

An important leap forward, according to the experts, took place when unicellular green algae appeared on the scene. These algae were capable of making photosynthesis, which takes advantage of water and CO_2 to synthesize carbohydrates and release free oxygen to the atmosphere. The combined effects of plate techtonichs and comet and asteroid bombardments had been building up the hydrosphere and the atmosphere of the planet, together with those primordial biochemical process, to rise the oxygen level in the atmosphere to its present level about two billion years ago.

Finally, after a series of catastrophic extinctions, no less than five in the last half billion years appeared the vertebrates. And, just around the time of the latest glaciations, men.

It is said that the human genome differs only in 1% from that a chimpanzee. But, there must be something else than the mere genome.

Men, "the trousered apes", are the only ones who smile, the only ones who have conquered fire (and much later electricity), who have cultivated the land, growing wheat, rice, corn…, who have tamed other animals for their own benefit.

The only ones who speak, write (in a hundred alphabets) make symbols, sing and dance and play guitar, organize great Olympiads and build great walls (like in China), large pyramids (like in Egypt) or astonishingly beautiful cathedrals (like in Europe).

The only ones that communicate with each other at the speed of light, that move from one place to another with supersonic speed; the only ones that make art (music, painting, literature); the only ones who make history; the only ones who make science (mathematics, physics, chemistry…); the only ones who have traveled to the Moon.

And the only ones who have traced back cosmic history to the Big Bang, approximately 13.7 billion years ago.

Bibliography

1. S. L. Jaki, "Science and Creation" (New York: University Press of America, 1990).
2. "National Geographic" 137 (May 1985) p. 662.
3. S. Weinberg, "The first three minutes" (New York: Bantam Books, 3rd printing, 1980).
4. J.C. Matter and J. Boslough, "The very first light" (London: Penguin Books, 1998).
5. H. Elsässer, "La Vía Láctea" en "Cosmología Astrofísica", Eds: J. A. Gonzalo, J. L. Sánchez Gómez, M. A. Alario (Madrid; Alianza Universidad, 1995).
6. Sir Isaac Newton, "Mathematical Principles of Natural Philosophy and Hys Sistem of the World" translated and edited by F. Cajori, p. 554 (University of California Press: Berkeley, 1934).
7. "Allgemeine Naturgeschichte and Theoriedes Himmels", Werke I, 360.

Chapter 11

From the Big Bang to the Present

The Universe as it appears now is basically homogeneous and isotropic, therefore with spherical symmetry; the constituent galaxies are moving radially away from each other with speeds proportional to the distances between them, with a local specific expansion rate given by $H_0 = \dot{R}_0 / R_0$; this fact together with the estimated age of the oldest stars in our galaxy (and those in other galaxies), which is of the order of t_0 (the time elapsed since the <u>Big-Bang</u>) allows us to determine a well defined origin in time for the cosmic expansion; the relative abundance of ^{4}He, and other light elements (which have a "primordial" origin) allows us to put a specific bound to the cosmic expansion rate at the time of the ^{4}He nucleosynthesis; finally, the present equivalent temperature (T_0) of the cosmic background radiation, combined with the ionization/formation temperature of atoms (T_{af}) taking place out of the preexisting plasma of nuclei (mostly H and ^{4}He) and electrons (a temperature which happens to coincide almost exactly with the temperature at which radiation and matter energy densities are equal (T_{eq})), allow the determination of the characteristic cosmic temperature (T_+) at which the first stars and galaxies began to exist. This temperature is directly related to the characteristic radius, (Schwarzschild radius) of a universe with a total finite mass M_u (i.e. $R_+ = 2\,GM_u/c^2$).

Consequently, the <u>Big-Bang</u> can be visualized going back in time (reversing the present galactic stampede) and reconstructing cosmic requirements, at least tentatively, to the cosmic conditions when the universe was much smaller, much more dense and hot that it is at present. Starting from the present phase of the expansion, in which the universe is formed by galaxies separated from each other by large distances, and is basically transparent to radiation, the cosmic equation of state which relates mass density,

$$\rho_m = M/(4\pi/3)R^3 \qquad (11.1)$$

and temperature, T, is given[1] by

$$\rho_m T^{-3} = \frac{\sigma}{c^2} T_{af} = 3.23 \times 10^{-32} \text{ g/cm}^3\text{K}^3, (t \geq t_{af}) \qquad (11.2)$$

where σ is the Stefan Boltzmann constant, c the velocity of light, $T_{af} =$ 3808 K, corresponding to a time $t_{af} = 1.47 \times 10^{13}$ seg $= 0.466 \times 10^6$ years.

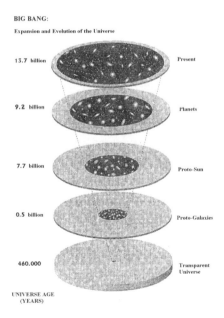

Fig. 11.1. Big-Bang.

At earlier times, $t \leq t_{af}$, i.e. when the universe was in its plasma phase the temperature was too high to allow the existence stable atoms. The eventual atoms formed furtively with temporarily trapped electrons by the nuclei of H (approximately 75%) and ^{4}He (approximately 25%) were ionized again very quickly. It is not very clear what was the effective equation of state to describe the cosmic evolution in this previous (plasma) phase. But it is not realistic to assume that cosmic expansion took place in the same way, and at the same rate, when the universe was very hot opaque plasma, and when it was a substantially cooler transparent gas, made up of atoms soon condensed into stars and galaxies far away from one another.

Previously the same equation of state applicable for the present transparent phase has been used customarily. But there are good reasons to think that for a plasma universe the pertinent equation of state,[1] instead of the one given above, is

$$\rho_m T^{-4} = \frac{\sigma}{c^2} = 8.47 \times 10^{-36} \text{ g/cm}^3\text{K}^4, (t \le t_{af}) \quad (11.3)$$

which describes well matter and energy expanding together at the same rate, while the universe remains opaque. In a transparent universe radiation expands more freely than matter. At present, $(\rho_{r0}/\rho_{m0}) \approx 10^{-5}$. And from now on, after some billion years, the ration (ρ_r/ρ_m) will be even smaller.

As it has been said, at $T_{af} = 3808$ K the ration was $(\rho_{raf}/\rho_{maf}) \approx 1$. This justifies the description of the present epoch as the matter dominated epoch $(t > t_{af})$. But the previous epoch $(t < t_{af})$ corresponding to the plasma universe, would be characterized[1] by an early period of equality between radiation energy density and matter energy density, rather than by radiation domination.

This would mean that at $t < t_{af}$ (i.e. at $T > T_{af}$) in the plasma (opaque) phase of cosmic expansion, matter and radiation would be expanding in unison. Radiation would undergo continuous "scattering" by the nuclei and electrons in the cosmic plasma, exercising simultaneously a substantial radiation pressure on the charged particles. This sustained momentum transfer from radiation to charged particles (condensed latter in stars and galaxies) would provide a natural physical explanation of the radial flight of present day galaxies from each other.

Below, we will use $\rho_m T^{-4} = \sigma/c^2$ as the equation of state for the plasma phase of the universe noting that the alternative equation $\rho_m T^{-3} = (\sigma/c^2)T_{af}$ (which is valid only at $t > t_{af}$) would give quantitative unacceptable results at nucleosynthesis.

As we have said, going back in time, the universe would become smaller and hotter as we approach the <u>Big-Bang</u>. Eventually, we would approach a (physically meaningful) minimum time,

$$t_L = h / M_u c^2 = 1.4 \times 10^{-103} \sec \ (M_u \approx 5 \times 10^{55} g) \qquad (11.4)$$

beyond which Heisenberg's uncertainty principle would preclude further meaningful inquiry. We can call this time Lemaitre's time, lacking a better name. At this time the cosmic temperature would be given by

$$T_L = M_u c^2 / 2.8 k_B = 1.1 \times 10^{92} K \qquad (11.5)$$

and the cosmic radius by

$$R_L = h / M_u c = 4.4 \times 10^{-93} cm \qquad (11.6)$$

assuming that the same plasma equation of state ($\rho_m T^{-4} = \sigma / c^2$) is still applicable. This plasma equation of state reduces to

$$RT^{4/3} = \text{constant} \qquad (11.7)$$

and to

$$t = t_x (T_x / T)^2, \text{for } T_x = T_{af} >> T_+, t << t_{af} \qquad (11.8)$$

With this can follow cosmic evolution through the successive stages previous to atom formation.

At these infinitesimally small times after the Big Bang, the density of the cosmic sphere containing all the matter density of the universe was so high that surpassed by many orders of magnitude that of protons, neutrons, etc, which, therefore could not have individual existence as differentiated particles. At much lower temperatures, however, as it is well known, photons of sufficient energy can create particle-antiparticle pairs, and vice-versa. For instance, photons with energy E > 0.5 Mev can produce, under certain circumstances, electron-positron pairs, and vice-versa.

For a certain period of the cosmic expansion, at high enough temperature, particle-antiparticle pairs were constantly being created and annihilated. As the universe cooled, the characteristic energy of the photons making up the cosmic radiation background was becoming lower and lower, till reaching the time when they were below the threshold for pair creation. But we know that at present time the preponderance of matter over anti-matter is overwhelming. Antimatter is detected only in very special circumstances and in very small quantities. At present there is not a satisfactory physical theory to justify the asymmetry matter-antimatter. In case the original amount of antimatter would have evolved into some kind of "dark matter" (weakly interacting massive particles for instance), their total amount of mass have been the same as that of the ordinary material particles not many times higher. Some inflationary theories postulate ten or thirty times more "dark matter" than ordinary matter. Other postulate even higher amounts of "dark energy".

When Planck introduced in 1900 his famous theory of the energy quanta to explain the "black body" emission, he introduced also what[1] he called natural units of mass, length and time, built up exclusively on universal constants. The mass of what was called later a Planck monopole is

$$m_{pl} = (hc/G)^{1/2} = 2.17 \times 10^{-5} g \tag{11.9}$$

much larger than a proton mass ($m_p = 1.67 \times 10^{-24}$ g). Since to any particle of mass m we can assign a characteristic temperature $T = mc^2/2.8k_B$, the temperature corresponding to a Plank monopole would be

$$T_{pl} = m_{pl}c^2 / 2.8k_B = 1.26 \times 10^{32} K \tag{11.10}$$

billions of billions of times higher than the temperatures presently existent in the interior of stars.

Assuming that at temperatures lower than Planck's are valid the known physical laws and the same relativistic cosmic equations[1] (complemented by the equations of state given above) cosmic events develop as matter and energy expand and the universe cools down from the initial very high temperatures: baryon (proton and neutron) formation, electron formation, ^{4}He nucleosynthesis (and some other light elements), atom formation, at the moment at which the universe becomes transparent and light, the "first" light, begins to propagate freely in all directions. As time develops, when the temperature has gone down sufficiently, galaxy and star formation begins and the universe acquires its present characteristics.

Table 11.1. Summary of quantitatively cosmic events.

Époque	Mass (g)	Temperature (K)	Time (s)	Density (g/cm^3)
Planck	2.17×10^{-5}	1.26×10^{32}	1.16×10^{-44}	2×10^{93}
Baryon	1.67×10^{-24}	3.88×10^{12}	1.43×10^{-5}	2×10^{15}
Electron	9.10×10^{-28}	2.11×10^{9}	49	1.7×10^{2}
^{4}He synthesis	-	3.47×10^{8}	1.80×10^{3}	0.125
Atoms	-	3.80×10^{3}	1.47×10^{13}	1.8×10^{-21}

When the universe became transparent at the temperature of formation/ionization of atoms (hydrogen and helium) the primordial fluctuations in density and temperature seen for the first time by COBE were already in place. This occurred at a time $t_{af} = 1.47 \times 10^{13}$ sec = 4.6×10^5 years. At that time there were no stars and no galaxies.

In other words, a little less than five hundred thousand years after the Big Bang the universe went from its previous plasma phase to the present phase of neutral atoms. This happened at a time when a little less than the 13.7×10^9 years gone by since the moment of the Big Bang. Much later, the first generation of stars and the first galaxies, including our Via Lactea, would begin to from.

Using Einstein's cosmological equations[1] we get

$$t_+ = t_0 \frac{\sinh(y_+)\cosh(y_+) - y_+}{\sinh(y_0)\cosh(y_0) - y_0} \approx 0.53 \times 10^9 \text{ years} \qquad (11.11)$$

for the time of formation of the first stars and galaxies, which we identity as the Schwarzschild time, when the cosmic radius was the Schwarzschild radius corresponding to the total mass of the universe. This is the same as Jeans radius. This happened when $R = R_+$, i.e. when $y_+ = 0.8813$. See for instance original paper[4] published in "Acta Cosmologica" which takes $y_0 = 1.92$.

[Note that the number for y_0 is slightly modified in latter work which takes into account WMAP's first year data for H_0].

This means that keeping $t_0 = 13.7 \times 10^9$ years as the present "age" of the universe, the initial estimate of t_0, which, in principle, is not unreasonable.

Using on the other hand $M_u = M_+ = 5 \times 10^{55}$ g (which might be slightly subestimated), and the $t = t_+$ radius of the universe $R_+ = 2GM_+/c^2$ (Schwarzschild radius) we get

$$\rho_+ = \frac{M_+}{\frac{4\pi}{3}R_+^3} = 2.93 \times 10^{-29} \, g\,/\,cm^3 \tag{11.12}$$

as the density corresponding to the time of formation t_+ of the first stars and galaxies.

At times less than t_+, $R < R_+$ well differentiated subunits (fractions) of the total mass of the universe could not be gravitationally bound and could not exist as such. Even electrons could not exist at $t < t_e$ ($R < R_e$) and baryons at $t < t_b$ ($R < R_b$).

But once the first generation massive protostars were formed within the first galaxies, they did burnt out quickly, resulting in "novae" and "supernovae" throwing in all directions vast quantities of heavy nuclei which would quickly become part of cosmic dust for stars and planetary systems of second and even third generation star, as our Sun. In one of these stars, under very special conditions, the right proportion and total amount of carbon, oxygen and nitrogen would be met to produce a hydrosphere and atmosphere with the potential for supporting life.

In Table 11.2 the quantitative data characterizing key moments of cosmic evolution from the time at which the universe became transparent to the present époque are given.

Table 11.2. Quantitative data of cosmic evolution.

Époque	Temperature (K)	Time (s)	Density (g/cm^3)
Transparent Universe	3.8×10^3	4.6×10^5	1.78×10^{-21}
Formation Via Lactea	30.4	0.53×10^9	2.93×10^{-29}
Present Époque	2.726	13.7×10^9	6.54×10^{-31}

[Note that the numbers for temperature, time and density corresponding to the formation of the Via Lactea are somewhat modified in later work which takes into account WMAP's data].

The story of the last thirteen billion years beings with the condensation of enormous cosmic masses of the order of 10^{44} g in galaxies. Among the galaxies then formed is our Via Lactea that contains about 10^{11} to 10^{12} stars and an amount of interstellar mass[5] (atoms, molecules, dust particles…) of the order of the ten percent of the mass contained within stars.

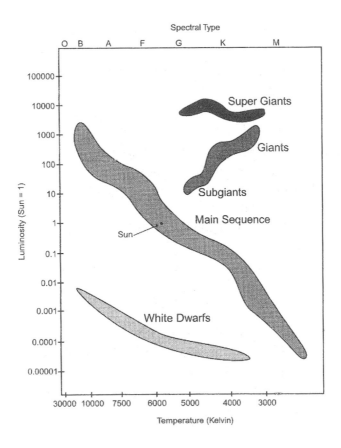

Fig. 11.2. Hertzsprung-Russel diagram.

The Via Lactea presents a multitude of stars of different generations. The Hertzsprung–Russell diagram (absolute luminosity as a function surface temperature) allows one to estimate the "age" of the star on the basis of the deviations of the star in question with respect to the line of the "main sequence". The stars with mass one hundred times that of the Sun appear located in the upper part of the diagram, and the theory of stellar structure and evolution allows one to assign them ages of just 10^7 years, while stars in the lower part may have present ages above 10^{10} years, but not much more, because they cannot exceed the age of the galaxy. Up to now, the Sun has deviated little from the "main sequence". Consequently its age is about 5.5×0^9 or less, which is compatible with the ages of the planets, estimated independently, which is about 4.5×0^9 years. The ages of the oldest stars in the Via Lactea, probably coeval with it is of the order of 13×0^9 year. The younger stars circulate in the galactic plane while the first and oldest generation of stars practically does not participate in the galactic rotation and move in plains which strongly deviate from the galactic plain.

Recent investigations of the characteristics of the galaxy center indicate that at its center our galaxy presents an Active Galactic Nucleus (AGN) similar to those detected in other galaxies. Our Milky Way, like Andromeda, is not a properly so called active galaxy, but might have been active in the past. The investigation of the center is difficult, among other things, because it is not accessible in the optical spectral range. In the infrared however, the central interstellar dust is much more transparent. It is commonly admitted that an AGN consists basically in a black hole, fed by an accretion disk, gravitational energy is converted to radiation energy with great efficiency, which explains the enormous luminic intensity of an AGN. From the rotation velocity of the masses concentrated in the center's proximity the mass of the black hole can be estimated as being about 2×0^6 times the solar mass. All this mass is concentrated within a region with radius of the order of 0.1 pc approximately.

It is known that the mass in our galaxy is greater than the "visible" mass, i.e. the mass concentrated in stars and cosmic dust. Some estimates of the extra mass in the galaxy give as much as several times the "visible" mass, but the question is not yet clear.

As we have said, the Sun is a second or third generation star formed by cosmic dust (which includes already a substantial proportion of heavy elements), and mainly hydrogen plus helium in the usual proportion. Probably it begun to radiate six to seven billion years after the formation of the galaxy. Since then, after about five billion years ago, the remaining primary fuel (H) is sufficient to keep it radiating at about the present rate for about another five billion years.

We know that Kepler's laws and Newton's gravitational theory describe well the motions of the planets in the solar system, Mercury, Venus, the Earth, Mars, Jupiter, Saturn, Uranus and Neptune, as well as those of Pluto an the comets (See Table 11.3)

Table 11.3. Solar system.

Planet	Mass (kg)	Orbit radius (Km)	Temperature (K)
Mercury	0.32×10^{24}	57.9×10^6	482
Venus	5.80×10^{24}	108.7×10^6	351
Earth	5.90×10^{24}	149.6×10^6	300
Mars	0.61×10^{24}	227×10^6	243
Jupiter	1900×10^{24}	778×10^6	131
Saturn	570×10^{24}	1428×10^6	97

However for question such as why the number of planets is such, no more and no less, why they have the mass they have, the mean orbital radius they have, the angular momentum they have, why all rotate in the same sense and do it in the same plain, there are no definite answers for the moment.

Some theories[6] predict the formation of planets as natural companions in the process of star formation. According to them, in parallel with the formation of the central mass of the star, smaller bodies made up of solid matter begin to form. After some time, according to this theory, sufficiently large fragments begin to act as condensation centers, resulting in size to "planete simals". This must have happened about 4.6×10^9 years ago. If the mass is much smaller than the mass of the Sun (Jupiter, the most massive planet is 50 times less massive than the Sun), the temperature at the center, where the pressure is higher, is still below

20×10^6 K, the minimum temperature needed to produce fusion, and the massive planet does not become a small and long lived star.

The masses of the planets in our solar system are very different from one another, as shown in Table 11.3. For instance, the Earth's mass is 318 times less than that of Jupiter. It is evident that in the interior of the Earth nuclear fusion is not possible. But, the heat generated by radioactive decay results in sufficient heat to produce volcanic activity associated with plate tectonics. At remote geological times volcanic activity was essential, together with asteroid and comet bombardment, to produce the oceans and the Earths atmosphere.

We live in the Earth-Moon System. The Moon has a diameter somewhat less than one third of the Earth's diameter, and a mass somewhat more than ten per cent of that of our planet. Viewed from the Earth, the size of the Moon is practically the same as that of the Sun, which has been essential to make possible the observation of solar eclipses, a fact of tremendous importance for the development of astronomy in antiquity, and especially in Greece. Thanks to the eclipses it was possible for Aristarchus, as already seen, to make the first estimates of the size and distance for the Moon and the Sun in terms of the radius of the Earth, established by Erathostenes somewhat earlier. The Moon is responsible for the tides which have had such a great importance to make viable the first "habitats" for marine life in the oceans and for the transition form them to firm land. It has been also essential as stabilizer for the rotation of the Earth at a convenient angle with the perpendicular to the plane of the ecliptical so as to make possible the four reasons and the possibility of annual harvests. The Moon has no atmosphere and has been exposed to heavy bombardment for many millions of years, acting as a complementary protective shield for the Earth.

About one billion years after its formation the Earth was probably a rocky sterile body (similar to Mars today) which gradually was evolving towards what it is today, with its oceans and continents, decorated by rivers, lakes and mountains, with abundant vegetation in large extensions of both hemispheres of the globe.

In the 20th century it was first conjectured by A. Wegener and then confirmed and carefully documented, that about 225 million years all continents were united in a single large continent (Pangea) which broke away gradually in the present day continents.

The fossil record testifies about the tremendous variety of living beings which have populated the continents and the oceans all along three billion years. Now there are at least 20,000 different species of fishes and myriads of other vertebrates culminating in mammals, apes and finally men.

The DNA of the nuclei of the simplest living cell directs in such efficient way the ensambling of the nucleotides and the aminoacids in its vital proteins that it appears extraordinarily unlikely that it was due to a succession of random gradual changes. Gradual evolution or intelligent design? Both may actually coexist to a certain extent. But, as A.R. Wallace, co-proponent with Ch. Darwin of the "natural selection" as the mechanism driving important changes in the living systems, three "discontinuous" steps in the history of life on our planet, (a) when the first cell was created, (b) when the animal kingdom departed from the vegetable kingdom, and (c) when man was created, must be singled out. Molecular Biology is making great advances[7] in describing details of the evolutionary mechanisms operative in living systems from the protozoans to the higher mammals.

Fig. 11.3. Blue Planet.

The first human organized societies, the first cities, appear only ten thousand years ago. Somewhat earlier men had begun to cultivate the land and to make use of domestic animals. About five thousand years ago[8] they begun to invent rudimentary writing and to work with clay and glass. Soon they began to use bronze and to make the first calendars. The first year recorded[9] by the Egyptian calendar of 360 days and 12 months of 30 days each is 4241 BC. About that time, more than six thousand years ago, in Egypt, in Babylon, in India, in China men were making the first systematic astronomical observations.

In Egypt, the use of papyrus is discovered an the first libraries begin. Egyptian ships transport gold. In China, the equinoxes and solstices are recorded. Not much later Abraham leaves Ur of Chaldea (the year 2100 BC) for the promised land.

Around the year 520 AC, in the lands surrounding the Mediterranean, i.e. the lands of the former Roman Empire calendars change the dating from year of foundation of Rome to the year estimated of Christ's birth, the first Christmas.

1968 years (more or less) after Christ birth, the first American astronauts set foot on the Moon.

Bibliography

1. J. A. Gonzalo and M. I. Marques Arxiv: Astro-ph/990561. 13 May 1999.
2. Max Planck "Physickalische Albandlngen und Vorträge" 1; 666 (Braunschweig: Fiedr-Viewg & Sons, 1958).
3. J. A. Gonzalo and M. I. Marques, see Ref.1 above.
4. N. Cereceda, G. Lifante an J. A. Gonzalo, Acta Cosmologica (Univ. Jagellonica) Fasciculus XXIV-2 (1998).
5. H. Elsässer, "La Vía Láctea" in "Cosmología Astrofísica", Eds: J. A. Gonzalo, J. L. Sánchez Gómez, M. A. Alario (Madrid; Alianza Universidad, 1995).
6. R. Jastrow, "Red Giants and White Dwarfs" (Warner Books: New York, 1979).
7. Michael J. Denton, "Natures Destiny: How the Laws of Biology reveral Purpose in the Universe" (The Free Press: New York, 1998).
8. R. E. Hummel, "Understanding Materials Science" (Springer: New York, 1998).
9. B. Grun, "The Timetables of History" (Simon and Schuster: New York, 1963).

Chapter 12

Astrophysical Cosmology
Around Year 2000 AC

12.1. The COBE Project

In the 18th of November 1989, a satellite with very special characteristic, the COBE — Cosmic Background Explorer — was launched in a Delta rocket. Its mission was to measure, with unprecedented precision the characteristics of the cosmic background radiation predicted by Alpher and Herman in 1948 and detected by Penzias and Wilson in 1964. In 1974, fifteen years before COBE's launching, John C. Mather having just completed his graduate studies in physics at the University of California, Berkeley, responded to a call for proposals by NASA to be carried out through of a satellite to be put in a orbit by means of a Scout or Delta rocket. The first schematic diagram for the project already included three experiments: the "Differential Background Radiometer" (DMR). The "Diffuse Infrared Background Experiment" (DIRBE) and the "Far Infrared Absolute Spectrometer" (FIRAS). Fifteen years are many years and during them COBE was to go through many vicissitudes, including the "Challenger" disaster of 1986, which made to more than one ask himself in political circles and the media whether NASA was really necessary.

In 1976–77, experiments by D. Wilkinson at Princeton, soon thereafter confirmed by Miller and Smoot, had detected a "dipolar" anisotropy in the CBR related to the motion of the Earth in a given direction with respect to the geometrical center of the expanding sphere of quasi-isotropic radiation. Two groups at Princeton and Florence in

1981 had made experiments in aerostatic balloons which seemed to indicate a certain "quadrupolar" asymmetry unrelated to the Earth's motion. In 1987 a Nagoya–Berkeley collaboration in an experiment carried out on a missile flying at 200 miles had reported a 10% excess over the characteristic blackbody spectrum in a certain spectral region.

John C. Mather recounts years latter COBE's launching[1]: "Top officials from NASA headquarters were there. As the final countdown commenced, Len Fisk stood next to the McDonnell Douglas engineer who was to give the Delta its final command to launch. Fisk had authority to abort the procedure up to the last second if necessary. At approximately 6:34 a.m. the sky lit up and the rocket, slowly and first, in eerie silence begun lifting off the pod, Chuck Bennet, seeing the flash of light, thought at first that the rocket had exploded. Within seconds it was racing faster than the speed of sound, its contrails winding dramatically around a quarter Moon.

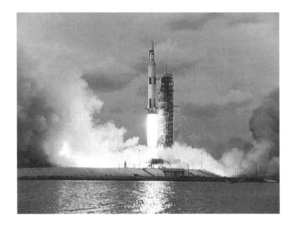

Fig. 12.1. The COBE launch.

"It was a gorgeous, joyful sight", Chuck recalled. Many in the crowd gasped as, a little more than a minute after launch, six solid fuel rockets were ejected at an altitude of more than 10 miles. But not Dennis McCarthy. He had been unable to watch (See launch illustration…).

"Within an hour the COBE was a satellite, circling the planet in an almost perfectly circular north-south orbit with an altitude ranging from 899.3 kilometers to 900.5 kilometers that followed the moving dividing line between night and day, making a complete circuit every 103 minutes. Dennis McCarthy was elated..."

"...We had to get back to Goddard and find out what kind of data COBE would generate about the beginning of the universe in the hope that the knowledge we would gain would justify its $160 million cost (not counting Civil Service salaries or the rocket) so far. The total cost, counting everything, was probably about $350 or $400 millions".

The DIRBE near infrared map of the sky, was taken with radiation at three wavelengths (25, 60 and 150 micrometers) which are longer than that of visible light. The photo depicts the heart of the Via Lactea, which cannot be seen with visible light.

Further maps of our galaxy taken with DIRBE show it is slightly distorted and that its center is not symmetric. It did show also that the cloud of interplanetary dust in the solar system is not centered in the Sun, but displaced some million miles from it, due to the gravitational influence of Jupiter and Saturn. DIRBE's principal investigator was Dr. Micke Hauser.

The FIRAS spectrum of the cosmic microwave background was one of the most spectacular. This beautiful result, showing a perfect agreement between the data points and the solid line giving Planck's radiation law for a blackbody at a temperature of 2.767 ± 0.0 K was one of the most hoped for result produced by COBE.

John C. Mather says in his book[2]: "The day of our presentation (January 13, 1990) arrived... I was somewhat annoyed, thinking that by the time of our presentation — the last session of the meeting — everybody would have left... Mike Hauser, George (Smoot), and I were dressed similarly for our presentation in what must have looked like the COBE uniform — blue blazers and khaki pants — even though I was certain we would be speaking to an empty hall... As we walked into the large room, which I later learned could hold as many as two thousand people, I was astonished: The room was filled to overflowing. Nancy Bogges stood up and gave a summary overview of the COBE project".

"Then I took the podium. After describing the instrument's principle of operation, I displayed a graph of the spectrum of the cosmic background radiation as related by FIRAS. "Here is our spectrum", I said "The little boxes are the points we measured and here is the blackbody curve going through them. As you can see, all over points lie on the curve". The theoretical blackbody curve predicted how the blackbody radiation should look if it had truly originated in the Big Bang".

"There was a moment of silence as the other scientists there grasped the meaning of the data curve. Then the audience rose, breaking into spontaneous ovation... For me this was a moment of supreme epiphany. The COBE project had been such a part of my life for so long: all the proposals and reports, NASA's bureaucratic jungle, the scientific problems, the engineering challenges of reconfiguring the space craft after the Challenger disaster, all the personal difficulties. I may have begun to take the result for granted by now. Obviously nobody else had..."

COBE's DMR one-year map showed for the first time the minute anisotropies signaling the first proto-galaxies. They were shown by George Smoot, DMR principal investigator, at another meeting, the April Meeting of the American Physical Society, held near Washington in April 20–23, 1992. George stood up and made and excellent presentation.[3,4] He said "We have a quadrupole", and explained that it was only part of the irregularities observed. "There irregularities, unlike the dipole, arise not from our motion in space, but from the cosmos itself. They are the oldest and largest structures in the universe. At the press conference that followed, with the room filled with cameras, bright television lights, microphones and more than 100 reporters, George Smoot explained: "These are the primordial seeds of modern day structures such as galaxies, clusters of galaxies and so on. Not only that, but they represent huge ripples in the fabric of space-time left from the creation period".

"If you're religious, it's like looking at God". This remark received furtively more attention from the media than anything else. Even more perhaps than Stephen Hawking observation that finding the CBR anisotropies was "the scientific discovery of the century, if not of all time".

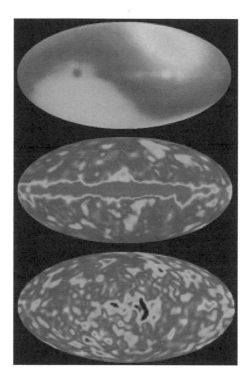

Fig. 12.2. Three full-sky made by the COBE satellite DMR.

12.2. The Hubble Space Telescope

The same October day of 1989 that COBE was being transferred from GSFC (Goddard Spatial Flight Center) in the Eastern coast of the US to its launching location in the Western coast, for pure coincidence, another great special probe, the HST (Hubble Space Telescope) was moved from California to Florida, to be tested and launched to space. They passed each other in route to their future respective launching place.[1]

Fig. 12.3. Hubble launch.

The HST was the result of a cooperation between NASA and ESA (European Space Agency). The possibility of a space telescope was enternained already at the end of the 40's, but HST was finally designed and constructed during the 70's and 80's. It was finally put in orbit by means of a Discovery space shuttle on April 24th, 1990. The original design[5] of the HST (1979) planned periodic return to the Earth every five years for their reconditioning and further re-launching to continue servicing for another 25 years. But in 1985, for various reasons, NASA discarded the idea of periodic returns by means of a shuttle. It was decided to keep in orbit the Hubble for a period of fifteen years and to adapt the design for this purpose. Missions to service in the orbit the HST were planned for December 1993, March 1997, mid 1999 and mid 2000.

Two months after having it in orbit, it was discovered that the main mirror of the telescope had a considerable spherical aberration. The light collected by the mirror was distributed in a diffuse zone instead of being focused in a specific point, due to a design error. To correct the distorted images, computer treatments were employed. They contributed to improve substantially those images. In this way the HST was able to observe candidates to black holes in the centers of some galaxies, expanding rings around "supernovae" etc. But, finally, a new wide field camera, with corrective optics incorporated, was substituted for the old one by astronauts in orbit. This way, the effects of optical aberration in the main mirror on several scientific instruments, in particular the camera and the spectrograph for faint objects and the high resolution Goddard spectrograph, were corrected.

The HST was a reflectant telescope with a length of 13.3 m and a diameter of 4.3 m, weighting approximately 11,000 kg. It was kept in orbit at 600 km over the Earth's surface which, being relatively closed to the atmosphere, was not a geostationary orbit and needed a little pushing up every now and then to compensate the descent, approximately one meter per year. The focal length was close to 58 m and the focal ratio f/24. The HST could make observations in the visible, in the near ultraviolet and in the infrared, covering a range of four orders of magnitude in wavelength (from 11.500 m to 1 mm).

The space telescope included three cameras, three spectrographs and two precision sensors. Due to its working above the atmosphere, these instruments produced high resolution images of astronomical objects of a quality very superior to those provided by the largest terrestrial telescopes, affected inevitably by the thermal fluctuations of the atmosphere, which did modify in a random way the optical path of the light coming from far away galaxies and stars through high vacuum in the intergalactic or interstellar space. The human eye can resolve two points separated 1' (one arc minute) from each other, a terrestrial telescope can resolve 1'' (one arc second), and the HST 0,1'', i.e. ten times better than the terrestrial telescopes.

During the HST operations, the fine guide sensors (FGS) are kept aligned with the reference stars to keep the telescope pointing exactly to the chosen objective.

The data and the operating instructions for the satellite are transmitted through a system of auxiliary satellites (TDRS) with control base on ground at White Sands (New Mexico). The HST can store up to 24 h. of commands in its on board computers, which allows it to retransmit them later when no TDRS is communicating with the HST. Only 10% of all projects proposed by research groups around the world did receive observation time at the Hubble.

The main scientific instruments on board of the HST are:

<u>WF/PC2</u> (Wide Field Planetary Camera 2). It contains three wide fields in L and a high resolution camera to observe the planets.

<u>FOC</u> (Faint Objects Camera). It includes two complete detecting systems with image intensifying tubes (105 times more than the light received). Stellar objects with brightness higher than that corresponding to magnitude 21 must be alternated with filters to avoid saturation of the camera.

<u>FOS</u> (Faint Objects Spectrograph) and HRS (High Resolution Spectrograph). The FOS gives information on the chemical composition, temperature, radial and rotational velocity, magnetic fields, etc of celestial objects. The FOS analyzes much fainter objects in a wider spectral range (UV of 1150 Å to IR of 8000 Å) than the HRS. But the latter is able to analyze in detail very fine spectral lines with a maximum resolution of one part in 10^5, and can detect rapid changes in intensity in times of the order of 100 milliseconds.

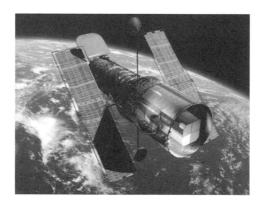

Fig. 12.4. Lateral view of the telescope.

Two solar panels of 2.4×12.1 m^2 supplied the electrical power needed (4.500 kW) to keep working the two computers on board and to recharge the six Niquel-Hydrogen batteries providing the electrical energy needed for the telescope itself during the 25 minutes in each orbit during which it remained in the shadow. A complete orbit lasted 95 min. To avoid the deleterious effect of the gasses emited by combustion motors, which would end up damaging the optics of the telescope, the spacial orientation of the HST was realized by means of three flywheels operated electrically. The system takes care of all the maneuvers required for reorientation and position control of the telescope. A set of gyroscope sends all data required to the control system. In the last mission performed before year 2000 three of the six gyroscopes, which had been giving problems, were substituted, keeping all the time the number of operative gyroscopes above the minimum for a perfect position control of the space telescope. The mission STS-61, carried out by the "Endevour" in December 1993 recovered most of the capacity lost due to the construction failure of the main mirror.

The STCcI (Space Telescope Science Institute) of the John Hopkins University (Baltimore), operated by NASA, is the one responsible for directing and co-ordinating all scientific operations of the HST. Once the main observation plan is realized the STScI passes control of the satellite to the Goddard Space Flight Center. The instructions for the orientation and other commands are transmitted to computers on board of the satellite which maintain communication with Earth seven times per day to confirm that everything is all right.

The key missions of the HST were the following:

- Precise determination of great cosmic distances (R) to determine with higher precision the Hubble parameter ($H_0 = \dot{R}_0/R_0$). Distances (are determined by means of variable stars of the Cefid type up to 60 million light years (Terrestrial telescopes are able to go only up to 12 million light years). Through these measurements the HST has reduced considerably the uncertainties leaving the value of Hubble's parameter in $H_0 = 65 \pm 10$ km/sec/Mpc.

- This means that a galaxy located at a distance of one million light years from us is going away at a speed of about 50,000 km/hour. A galaxy at twice this distance is receding at twice the speed. [We may point out that for very very long distances strict proportionality is no longer to be automatically expected].

- Studies of absorption spectra of "quasars" (quasi-stellar objects), very distant protogallaxies with extraordinarily luminous nuclei, which must have evolved considerably since the light reaching us now was emitted. Measurements of Deuterium and Helium abundance in regions of the universe very far from our galaxy. Acquisition of valuable information on the primordial nucleosynthesis.

- Investigations of the planets of our solar system through WF/PC2 which complement previous interplanetary expeditions (Viking, Galileo, etc) to investigate individual planets. The HST has checked the variable behavior of the planetary atmospheres of Jupiter, Mars, etc, in their path around the Sun.

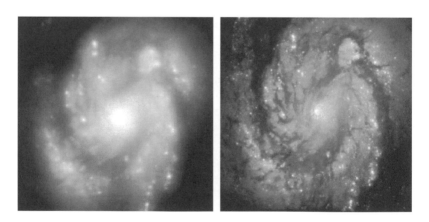

Fig. 12.5. Galaxy NGC-1068 before and after the repair.

- Confirmation of the existence of supermasive black holes. From the times it was demonstrated the finiteness of the speed of light (Roemer in 1649, Fizeau in 1849...) it was conjectured that gravitation could have some effect on it. John Mitchell (Cambridge) in the 18th century noted that a massive star sufficiently compact could produce a gravitational field so intense that any light emitted would be attracted back. But only after Einstein's general relativity, which allowed a consistent treatment of gravitational fields, firm grounds were set forth to discuss rigorously black holes. Karl Schwarzchild in 1916 established that the gravitational frontier (event horizon) for a very concentrated mass M was given by $R_{Sch} = 2GM/c^2$, where $G = 6.672 \cdot 10^{-8}$ cmg^{-1}s^{-2} is Newton gravitational constant, and $c = 2.997 \cdot 10^{10}$ cm/s the velocity of light.

If a star is consuming its hydrogen fuel producing first helium, then its helium producing carbon, then oxygen... and finally iron, the fussion reactions are the ones liberation radiant energy, which counters the gravitational attraction. Then a collapse takes place followed by a phase of equilibrium in which the repulsion between particles (Pauli's principle) plays the role of radiation, as noted by Chandrasekhar (1929). But this has also a limit. When the star has a mass two and a half times that of the Sun, Pauli's repulsion is insufficient to avoid collapse. Then, in some cases, an explosion is produced (nova, supernova). Landau proposed a final stage which he called neutron star. Later Oppenheimer (1939) proposed that a supermassive star would end up in a "black hole". Only half a century latter the first observational evidences of such black holes begun to be possible.

The accompanying Table includes some black holes catalogued originally by Hubble's Telescope.

Table 12.1. The Spacial Mission Hipparcos.

Galaxy (Constelation)	Distance	Mass (times M_s)
Via Lactea (-)	28×10^3 ly	2×10^6
NGC 224 (Andromeda)	2.03×10^6 ly	30×10^6
NGC 3115 (Sexant)	27×10^6 ly	2×10^6
NGC 4261 (La Virgen)	90×10^6 ly	400×10^6

12.3. The Spacial Mission Hipparcos

The Spacial Satellite Hipparcos[6] (High Precission Parallax Collecting Satellite) was launched to space from the French Guyana on board of an Ariane rocket by the ESA (European Space Agency) in August 1989. The acronym for the satellite was designed to honor Hipparcos of Nicea (190 to 120 B.C.) the great Greek astronomer who made a great catalogue of stars visible in clear nights to the nude eye. He was the one who noting some differences with the oldest catalogues elaborated by babylonian astronomers coined the term "nova" to designate stars not previously registered. He also recognized the precession of the equinoxes every 26,000 years which was identified by some Greek philosophers as the "Great Year" after whose completion everything would repeat again and again.

The European space satellite was provided with instruments of unprecedent precission to investigate the precise locations of stars in our galaxy which would trigger spectacular astronomical advances.

Hipparcos completed its ambitious program of measurements in 1993. The phase of analysis of the data was completed in 1998, with the precise identification and positioning of more than one hundred thousand stars. These results were subsequently published in 17 volumes[7] in a very complete catalogue containing the positions, distances, motions and many other properties of myriad stars in our galaxy.

Fig. 12.6. Satellite Hipparcos before being launched in 1989.

Stars differ from one another in their mass, their temperature, their chemical make up, their rotation velocity, their age, etc.

An adequate knowledge of their composition and structure suffices to ascertain how the star was born, how is evolving and how it will die, after having consumed all the available nuclear fuel. The physics of the stars is determined by the nuclear reactions inside, as well as by the heat radiation and convection which takes place in its interior under strong gravitational forces. And a realistic physical model should make predictions of the correct luminosity, surface temperature and other observable physical properties of the star. Before the Hipparcos mission, only a few dozen stars were characterized with a precission above 20%. After Hipparcos the number of well characterized stars multiplied enormously.

The precise determination of the stellar distances is not an easy task. The orbital motion of the Earth around the Sun, as it is well known, result in a small measurable stellar parallax which allows the determination of the distance to the star in question (See Figure 12.7). Hipparcos has measured this parallax with unprecedent precision for tens of thousands of stars distributed all throughout the galaxy. On the other hand, the positional distributions and the trajectories of the constituent stars of our galaxy, of the order of 10^{11} to 10^{12}, are related through their gravity and their formation history. In other words the positioning and the motion of a star keep the secret of its initial formation and subsequent evolution. This justifies the high cost of the astrometric spatial mission carried out by Hipparcos.

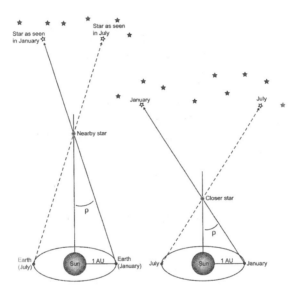

Fig. 12.7. Estellar parallax.

The catalogues prepared contain more than 20,000 stars with their distance to us within two to three thousand lightyears, determined with a precision higher than 10%, and about 30,000 additional stars determined with a precission better than 20%. This positional information is complemented with an information incomparably better than that available up to now on the velocities. Through photometric measurements, the existence of thousands of new variable stars has been demonstrated. This gives very valuable information for the characterization of the temperatures and other physical properties of the stars.

An emergent new branch of astrophysics is what is known as "astro-seismology". It studies the rhythmic seismic pulsations of stars, analogous to that being observed for some time in the Sun.[8] It has been observed that the Sun present characteristics internal modes of oscillation, randomly excited, resulting in measurable fluctuations with amplitude detectable by means of precission photometry, to levels of one part in a million. Each oscilating mode in a star is characterized by a radial number (n) and an angular number (l) as well as by a specific frequency determined by the internal conditions of the star. The mode frequencies of the lowest order depend on the velocity of sound through the stellar medium and through its surface: by means of appropriate stellar modes it is possible to obtain the absolute luminosity of the star. The direct comparison of this luminosity with the apparent magnitude of the star allows one to estimate the distance, which itself can be compared with the trigonometric distance (Hipparcos parallax), which is measured independently.

Till recently the most complete catalogue of stars of our galaxies was the "Catalogue of Nearby Stars" compiled with data taken along the past one hundred years and kept to date by the group of W. Gliese (U. of Heidelberg) since 1957. It contained only about two thousand galaxies in our Via Lactea and was considered as a representative sample. Today the extraordinary stereoscopic vision of our galaxy provided by Hipparcos has changed completely the panorama. Within a radius of 25 parsecs of the Sun, 200 "new" stars have been found. Several hundred stars previously counted within this radius are known today to be much further away. It has been checked that there are less stars in the main sequence of the Hertzsprung Russell diagram (see below). And 37 new binary stars have been contabilized in our vicinity, including a probably "brown dwarf" companion of the star Gliese 433. The density of mass in our close vicinity has been reduced to 0.039 solar masses per cubic parsec, much less than previously through. This reduces considerably the possible amount of "dark matter" contained in the galactic disk.

Careful investigations of the matter density throughout the galaxy lead to 0.076 ± 0.015 solar masses per cubic parsec, equivalent to $1.23 \cdot 10^{-24}$ g/cm^3, which is less that the estimated density for visible matter until very recently, not counting "dark mass", which, for the moment can only be estimated with considerable imprecision. On the other hand the stars at the outside of the disc rotate in unison with the stars in the inside, which would require an external "halo" of dark matter to avoid a "keplerian" distribution of velocities.

Our Via Lactea is a spiral galaxy whose disk rotates around an axis perpendicular to the disk through its center with a period of 250 million years. Besides this general rotation, the stars have proper motions, which reflect their gravitational history, having been perturbed by other individual stars or by giant molecular clouds moving through their vicinity.

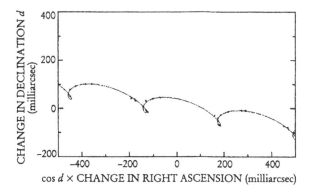

Fig. 12.8. The trajectory of 47 Ursa Majoris viewed on the background of the distant "fixed" stars.

In the accompanying (Fig. 12.8) we see the trajectory of 47 Ursa Majoris viewed on the background of the distant "fixed" stars. The curve represents the trajectory calculated by adjusting it to the Hipparcos measurements. The change in declination (d) in millisecs vs. the change in right ascension (times cos d in milliseconds gives the amplitude of the oscillations measured during three years. Small periodic Doppler displacements of the light coming from the star suggest that it has a great planet, about seven times the mass of Jupiter, circulating it.

Hipparcos has produced also very precise results on the "open" group Hyades which eliminate previous uncertainties. This group or con-glomerate of stars, relatively close to the Via Lactea (some 40 parsecs) has played an important role in the definition of the scale of cosmic distances. Measurements performed from Hipparcos, at distances for which the parallax method is very precise, have given us very good distances to some 200 stars of the Hyades group, as well as other very valuable information on the internal structure and on the motions of the stars, including the disintegration processes in the conglomerate and the masses segregation in it, which, being an "open" cluster, contrasts markedly with what is observed in "globular" clusters, formed by stars much older, probably coeval with the galaxy of which they form part.

The He abundance in a conglomerate of stars is a very important factor to characterize its evolution, but it is not easy to measure. The Hertzsprung Russell diagram consists in the representation of points giving a star luminosity vs its temperature (color). Hipparcos has produced the most spectacular H-R diagram produced up to date, including 40,000 stars. Using stellar evolutionary models, it has been possible to determine the age of the Hyades group in 0.625 ± 0.050 billion years, and the distance to its center of mass has been determined with a precision better that 1%, establishing unequivocally the first stretch of cosmological distances needed for a precise determination of the Hubble parameter, $H_0 = \dot{R}_0/R_0$.

An important class of stars, the variable Cepheids, has been playing a decisive role in the quantitative determination of the expansion speed of the galaxies through the universe.

The Cepheids are pulsating stars (the Polaris is a pulsating star) which have pulsation period, going from days to weeks, correlated with their mean absolute luminosity. If for some nearby Cepheid the luminosity has been determined with good precision, and therefore its distance to us, this is sufficient to fix the distances of farther away sufficiently luminous Cepheids. Measurements made from Hipparcos on the Cepheids of the Great Magellanis Cloud, very close to the Via Lactea, complemented by other measurements, have been instrumental in correcting upwards the extragalactic distances in about 10%, and consequently, decreasing the Hubble parameter, leaving it in

$$H_0 = 65 \pm 10 \text{ Km/secMpc.}$$

On the other hand, measurements realized from Hipparcos to date "globular" clusters, formed by the oldest stars in our galaxy, have resulted in an "age" for the Via Lactea, a typical spiral galaxy formed shortly after the universe became transparent (atom formation), in around

$$t_0 = 14 \pm 3 \text{ billion years.}$$

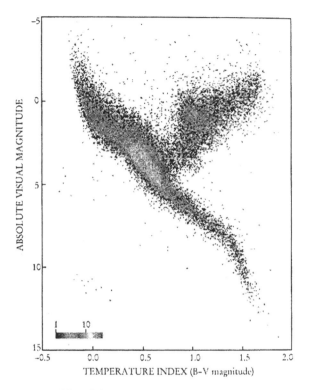

Fig. 12.9. Herzprung–Russel diagram.

Therefore, the data taken from Hipparcos have contributed in a decisive manner to make compatible the quantitative values for H_0 and t_0, which, previously, based upon erroneous evaluations, seemed to indicate an "age" of the universe ($\sim H_0^{-1}$) less than the age of the oldest stars in our galaxy.

For the first twenty years of the present millennium, about 200 years after the great Greek astronomer Hipparcos, a new mission of the European Space Agency is measuring the positions and velocities (in three dimensions) of hundred of thousands stars in our galaxy with a precision of a few tenths of arc microseconds.

Bibliography

1. J. C. Mather and J. Boslough, "The very First Light" (London: Penguin Books 1998).
2. Ibid., p. 233.
3. Ibid., p. 250.
4. G. F. Smoot and Keay Davison, "Wrinkles in Time" (London: Little, Brown & Co., 1993), Chapter 14.
5. E. Navarrete, I. Encimas, J. Perales: "El Hubble y los Agujeros Negros" (UAM: Trabajo TCR, 2000).
6. M. Perryman, "Physics Today" pp. 38–43 (June 1998).
7. ESA: "The 'Hypparcos' and 'Tycho' Catalogues", SP. 1200, in 17 vols. (1997); See also http://astro.estec.esa.nl/Hipparcos.
8. "Physics Today" p. 19 (may 1995).

**The Microwave Anisotropy Probe
(Singapore/Madrid, 2005)**

Chapter 13

The Report of the WMAP's First Year Observation in the NY Times: 02/12/2003

The WMAP (Microwave Anisotropy Probe) was launched in June 2001 in a Delta rocket, the same type of rocket in which the COBE had been launched in orbit in November 1989. The success of the COBE mission had established beautifully the blackbody character of the microwave background radiation and, in so doing, had detected unambiguously for the first time the small anisotropics which signal the formation of the first protogalaxies. The WMAP's results (rechristened with the initial W to the memory of Princeton University astrophysicists David Wilkinson, recently deceased, who was a founding member of the partnership between Princeton and NASA's Goddard Space Flight Centre) confirmed and complemented COBE's findings with unprecedented accuracy.

The Delta rocket did carry the WMAP satellite on a journey, which, within a few weeks, brought it to the vicinity of the L2 Lagrange point,[1] a special point 1.5 million km antisunward of Earth. From that point of observation, remote and unobstructed, the instrument could map the sky continuously, with an unprecedented precision, measure the faint departures of the CBR from perfect anisotropy (Fig. 13.1), and detect (something which COBE was unable to do) the still fainter polarization carried by the quasi-isotropic cosmic background radiation. As shown below, this small polarization gave unexpected information on the first massive stars, about one hundred times the sun mass, which had formed in a relatively short time after the universe became transparent, and after the cosmic plasma became a gas of neutral atoms, sometime before they begun to become gravitationally bound.

Fig. 13.1. All sky map of microkelvin departures from isotropy of the 2.725 K cosmic microwave background, as given by WMAP (angular resolution about 12 arcminutes).

In February 11, 2003, the expected report of WMAP's first full year observations, was made public in the form of 13 separate preprints, full of data, accompanied by a wealth of preliminary analysis. The following day, The New York Times outlined the story: "For Astronomers, Big Bang Confirmation"[2]. The report pointed out right away that the task was to understand the dark stuff ("dark matter" and "dark energy") that "apparently makes up 96 percent of everything, and to investigate what happened in the Big Bang that gave birth to it all".

Cosmologists, according to the NY Times report, "do not know what dark energy is". They do not know either, in spite of a wealth of potential candidates, what dark matter is made of. One leading candidate, according to prestigious theoretical cosmologists, is the force associated with the cosmological constant, which Einstein introduced as a fudge factor in an attempt to keep the universe from collapsing, and later disavowed. Alternative proposals include a force field named "quinta essence".

But apparently, cosmologists agree, the analysis performed has not solved the dark energy problem. Dr. Sperger, from WMAP's team said its data seemed to favour Einstein's fudge factor but the whole thing remains, however, highly speculative.

The cosmic microwaves represent a snapshot of the universe as it was cooling through the temperature at which atoms begun to form from electrons and protons (and from a non negligible amount of a-particles, ^{4}He nuclei), at a time of about 380,000 years after the Big Bang.

After COBE had confirmed, in 1992, that tiny lumps were present in the almost isotropic cosmic background radiation, a series of smaller experiments investigating more closely the lumps had concluded that the geometry of the universe was flat (Euclidean), but they only glimpsed at small portions of sky for limited times. The new satellite, however, was scanning the whole sky every six months. The new map of the sky given in Fig. 13.1 was based upon the first year worth of data, but the satellite was designed to operate for four years. Improved accuracy should be expected after the completion of eight full scans of the entire sky or more.

The satellite's instruments were capable to measure, like a pair of Polaroid sunglasses, the polarization of the microwave radiation, in addition to measuring its brightness with an unprecedented precision and angular resolution. Those measurements were crucial to determine the era of formation of the first stars. In the same way, as light skipping off a lake's surface,[2] the electric and magnetic fields that constitute electromagnetic radiation bounce off cosmic ionised gas, showing definitive preference to vibrate in a particular plane, and therefore, to become polarized. Recently, astronomers had shown that polarization was imparted to the microwaves right at the moment when the first neutral atoms were formed. A new different polarization event was expected by the astronomers when stars were first formed by gravitational collapse out of the gas of neutral atoms. In stars, the reionisation of the hydrogen and helium atoms takes place again, a process which is taking place now at the surface of our Sun (a second or third generation star). As in the Sun's surface, the free electrons at the surface of the newly formed stars polarize cosmic radiation again leaving an imprint in the CMB.

A majority of astronomers expected that the first stars would have been formed about the time of formation of the most distant quasars around 800 million years ago. But it was a surprise for them to find, from WMAP's data, that the first stars (probably monsters 100 times as massive as the Sun) formed so much earlier, as Dr. Bennet, WMAP's Principal Investigator, did explain in his interview for the New York Times. These early and massive stars did burn rapidly and violently, transmuting primordial hydrogen and helium into heavy elements, like carbon and oxygen, and sending them out to space, to form future generations of stars, including, eventually, our Sun and its planetary system.

The WMAP's data could shed light on what might have been going on during and after the Big Bang. The theory of inflation, however, which has been dominant among theoretical cosmologists for more than two decades now, as Dr. Bennet noted, is often called a paradigm instead of a theory. A physical theory is always accountable to test and experimental checks. A paradigm, however, not so well defined, might have, in principle, such a degree of flexibility that new observations could be accommodated within it, without excessive difficulty. Inflation seems to be available now in such number of versions that one or other is likely to account for most conceivable observational findings Dr. Linde, from Stanford, inventor of one of the models ruled out by WMAP's data, said, according to the NY Times, that it was "great" that theories were getting "culled". Dr. Turner, another prominent cosmologist, was reported as saying: "This is the door to precision cosmology being opened. It's the first step in a long march".

The instrument's design[3] was tailored to improve by an order of magnitude the calibration of the previous probes used by the COBE, which were already extremely accurate. Due to its sensitivity and to its uninterrupted full-sky coverage the WMAP was able to measure, only with the first year's data, the first "acoustic" peak in the temperature of the cosmic background anisotropy, within very, small errors (see Fig. 13.2).

The data provided a precise numerical value for the time elapsed since the Big Bang, as given by

$$t_0 = 13.7 \pm 0.2 \text{ billion years} = (4.32 \pm 0.06) \times 10^{17} \text{ sec} \quad (13.1)$$

and a specific time for the occurrence of atom formation, (triggering cosmic transparency) as

$$t_{af} = 379 \pm 8 \text{ thousand years} = (1.19 \pm 0.02) \times 10^{13} \text{ sec} \quad (13.2)$$

WMAP's first year data provided also very accurate estimates of the Hubble constant at present time

$$H_0 = 71 \pm 4 \text{ Km/s.Mpc} = (2.31 \pm 0.13) \times 10^{-18} \sec^{-1} \qquad (13.3)$$

The dimensionless product $H_0 t_0$, obtained using the present value of H (which is time dependent) and t (obviously also time dependent) is given by

$$H_0 t_0 = 0.942$$

This dimensionless product corresponding to an open universe ($k < 0$) must have evolved from an early value $H_{af} t_{af} \approx 2/3$, at $t = t_{af}$, as previously noted, and is growing at present towards one. This is a clear indication that the universe was expanding faster at earlier times as viewed from the Earth now. But, as noted in the previous chapter, it would be not only confusing but wrong to say that the universe is actually accelerating now.

Not so long ago[4,5] the value of Hubble's constant was known with a precision no better than 25%. In fact some estimates, in which the expansion of nearby galaxies counted more, suggested $H_0 \ll 50$ Km/s.Mpc, while other estimates, perhaps relaying rather in farther away galaxies, favoured $H_0 \approx 100$ Km/s.Mpc. Having into account that the expression of Hubble's constant, as deduced from Einstein's cosmological equations (without the cosmological term) is strongly dependent on time (as it is the density parameter), both previous estimates, $H_0 \approx 50$ Km/s.Mpc (from nearby galaxies) and $H_0 \approx 70$ Km/s.Mpc (possibly from farther away galaxies), may not be as contradictory as they appear at first sight.

The slightly hot and cold spots in the all-sky map given by WMAP's first year data signal local regions at the beginning of the transparent epoch that show mass densities and energy densities only very slightly lower or slightly higher than the mean value. The expansions and contractions of such density fluctuations can be viewed as acoustic waves in the viscous elastic cosmic fluid in which, at the end of the plasma epoch, radiation pressure was competing against gravitational contraction. In fact, radiation pressure, dominant in the plasma epoch, can be viewed as the driving force for the cosmic expansion all the way since the Big Bang, at least since Planks epoch ($t \approx 10^{-44}$ s) and conceivably even earlier. The sound speed, limiting how fast hot or cold spots grow in the plasma epoch, was relatively close to the speed of light, about $c/\sqrt{3}$.

To extract the best numerical values of cosmic parameters from the CMB radiation it is convenient to obtain the angular power spectrum-of temperature fluctuations by decomposing the celestial map giving ΔT departures from the mean CMB temperature as a sum of spherical harmonics, $Y_{lm}(\theta,\phi)$. For a certain multipole l, the fluctuation power is given by the mean-square value of the expansion coefficients $a_{l,m}$, averaged over the $2l + 1$ values of the azimutal index m. There is no preferred direction in the CMB sky after subtracting the dipole contribution. This contribution is due to the displacement of the probe in cosmic space with respect to the reference system defined by the CMB itself. (The distribution of power in m varies randomly with the observer's position in the cosmos). In Fig. 13.2, it is shown the angular spectrum of the temperature fluctuations. The variance (in μK) is indicated by a shaded region, which is widest at small l. According to the report by Bertram Schwarzschild in "Physics Today", the extremely low quadrupole ($l = 2$) power in the spectrum, which is outside the variance of the calculated best fit as a function of l, might or might not have cosmological significance, but the accumulation of further data in the next years could improve sufficiently the statistics to provide an answer to this open

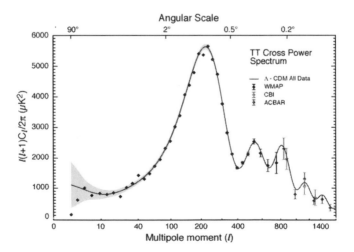

Fig.13.2. Angular power spectrum of temperature fluctuations in the cosmic microwave background given by WMAP.

question. The power spectrum peaks correspond to those modes (characterized by definite 1 values), which happened to be maximally either over or under-dense at the moment when the universe became transparent.

The main surprise provided by the first year of WMAP's observations, as noted, was the definitive excess of temperature-polarization cross-power (in μK^2), about 3 for 1 of order ten compared with an oscillating background about ±0.5 as a function of increasing 1 up to 1 ≈ 460 (see Fig. 13.3). This is attributed to the beginnings of cosmic gas reionization corresponding to the formation of the first stars, thought to have formed' about 200 million years after the original CMB temperature anisotropics were fixed in the microwave sky (at the time of atom formation) about 400 thousand years after the Big Bang.

Nearly twenty years ago,[4] M. Turner, G. Steigman and L. M. Krauss, did try to reconcile "theoretical prejudice" (in their own words), implying Alan Guth inflationary theory (with k = 0 as the cosmic curvature), with observational data, pointing out that the presence of mass smoothly distributed at scales >>199 Mpc could make compatible those data with

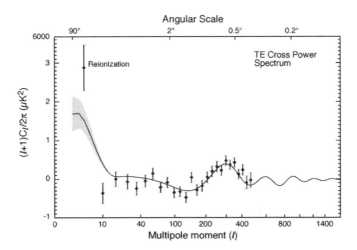

Fig. 13.3. Cross-power spectrum of correlation between CMB temperature fluctuation and polarization in the cosmic microwave background measured by WMAP. The point at the lowest multipole moment is attributed to the first stars, formed about 200 million years after the formation of the CMB according to WMAP's team.

$\Omega \approx 1$ (k = 0), either by means of relativistic particles, which might contribute a relativistic mass density $\Omega = 1 - \Omega_{NR} \gg \Omega_{NR}$, or else by reintroducing Λ (Einstein's cosmological constant) into the basic cosmic equation. It should be noted that Λ, by definition must be spatially constant. The contribution of a cosmological term into the density parameter would later be relabelled Ω_Λ.

A "Reference Frame" by Michael Turner in "Physics Today"[6] devoted to WMAP's findings is entitled: "Dark Energy: Just What Theorist Ordered". After referring to the problems for a flat universe considered most important at the 1990's he states: "To save a beautiful theory, theorists are willing to consider the implausible although not the impossible". (He does not explain why it should be considered so beautiful a theory which does not even attempt to respect energy conservation; a theory which is "the ultimate as a free lunch", according to Alan Guth). Then he reviews how Bondi, Gold and Hoyle used the cosmological constant to address the fact that the cosmic time back to the Big Bang appeared (then, half a century ago) to be less than the age of the Earth. Hints by the mid-1990 from the CMB anisotropy data could be taken as indications that the universe is flat. However, our author points out, "there was a problem: Λ_{CDM} (standing for "cold dark matter with a cosmological constant") also predicts accelerated expansion, and the first supernovae results did not yet show acceleration. With the discovery of cosmic speed up in 1998, Turner continues, "everything quickly fell into place".

Theoretical estimates of the cosmological constant (made obligatory in the authors view, by quantum mechanics, as a sum of zero point energies) give $\Omega_A = 10^{55} \gg 1$, somewhat embarrassing for theoretical physics, and to be left aside for the time being.

Let us stop here for a moment. As shown in the preceding Chapter, if counting distance (r) to supernovae from our vantage point implies accelerated cosmic expansion, counting the distance (R-r) from the Big Bang sphere (located behind the CMB radiation sphere, which is the right origin[5] for R), implies decelerated cosmic expansion.

After reviewing briefly the merits/demerits of the real or rhetorical problems which propelled cosmic inflation to the cosmologists attention (the monopole problem, the flatness problem and the horizon problem).

we will show how a time dependent $\Omega(t)$ is compatible with $\Omega(z = 0) = \Omega(to) \approx 0.04$, $\Omega(z = 1) = \Omega(t = 6.5 \times 10^9$ yrs$) \approx 0.08$ and $\Omega(t_{af}) \approx 0.98$, and how an evolving $H(t)$ is compatible with $H(z = 0) = H(t_0) \approx 67 \pm 4$ Km/s.Mpc, $H(z = 1) = H(t = 6.5 \times 10^9$ yrs$) \approx 135 \pm 8$ Km/s.Mpc and $H(t_{af}) \approx (4.96 \pm 0.30) \times 10^5$ Km/s.Mpc at atom formation.

The argument (2005) mentioned at the end of this Chapter explaining away accelerated expansion was wrong. For z in the interval $0.4 < z < 1.7$ the data favor accelerated expansion. For higher redshifts the question remains open (see comments in Chapter 18).

Bibliography

1. B. Schwarzschild, "Phys. Today", April 2003, p. 21.
2. "The New York Times", 02/12/2003 (Science Section).
3. B. Schwarzschild, "Phys. Today", April 2003, p. 21.
4. M. S. Turner, "Phys. Today", April 2003, p. 21.
5. M. S. Turner, "Phys. Today", April 2003, p. 21.
6. M. J. Rees, "Introductory Survey", pp. 3–20, in "Astrophysical Cosmology", Proceedings of the Study Week on "Cosmology and Fundamental Physics"; Vatican City, 1982).

The Medieval Roots of Contemporary Science
(Oviedo, 2007)

Chapter 14

Why not in China?

Chapters 14 to 17 below are based upon a lecture given by the author at Oviedo (Spain) on February 6th, 2007 at a Series of Conferences organized by Dr. Jorge J. Fernandez Sangrador from the Secretariant of Culture, Archbisopric of Oviedo.

14.1. Why not in China?

Some enlightened European freethinkers contemporary of Voltaire, having read extensive reports on China by Jesuit missionaries, were very favourably impressed with the Chinese excellence in ethics and moral philosophy as well as in the arts and the crafts but found them lacking in science.[1] The two-volume work of Father Louis Lecompte, SJ, mathematician to his Majesty Louis XIV of France, entitled "Nouveaux Mémoires sur l'état present de la Chine" (1696) was composed of fourteen long letters to various French dignitaries, civic and ecclesiastical, went through five editions in five years, and was quickly translated into English, German and Dutch. Well written, with numerous illustrations, and printed in a handy format, the book covered many topics on Chinese geography, politics, history, literature arts, crafts and also science. The science that Father Lecompte had in mind was Euclid's and Ptolemys[2] but not the kind of science which, with roots in medieval Christendom, through Copernicus, Galileo and Kepler, had just achieved full maturity in the "Principia of Newton".

In view of the considerable talents, speculative and practical, evidenced in the millenary Chinese culture, those enlightened Europeans, considering the meagre scientific level achieved there asked themselves: Why no science in China?

In other words, why the astronomy, the mathematics and, very especially, the physics in China was not comparable to the physical science in Europe at the time. The question raised was that of the historical origins of science, taking for science not a set of scattered partial achievements, more or less interesting, but not an organic system already in progress, as it was, already, European science by the closing years of the seventeen century. An organic system apparently destined to follow non-stop ascendant munch towards what we know today as contemporary science.

But, as shown by Stanley L. Jaki in his work "Science and Creation: From Eternal Cycles to an Oscillating Universe",[3] science in this sense, had not had a viable birth neither in one of the great civilizations of Antiquity nor in Babylon, nor in Egypt, nor in India, not even in the most brilliant of them, the Classical Greek civilization. The Greeks had had spectacular developments in geometry and notable achievements in astronomy, but not in physics. In physics, i.e. in the study of mater in motion, and of the interactions between material bodies, they had made practically no progress. Their scientific knowledge in mechanics, optics, thermodynamics was rudimentary, if any. And the creative period, scientifically speaking, among the Greeks, had been relatively short. Greek Science came soon to a stand still, and was followed by a clear decline.

The question to be asked by those European free-thinkers should have been another one: Why a genuine birth of science, a development susceptible of arriving, after some trial and error, to a final stage of self sustained scientific growth, had taken place in one civilization (our Western Christian Civilization) and only in it?

The question of how it began to develop in learned circles a firm and comprehensive scientific view of nature and its laws is certainly a momentous question, deserving a careful investigation and a balanced and objective response.

The earliest developments took place centuries ago in medieval Europe in a cultural matrix unequivocally Christian. There had been many other great civilizations in world history which could be proud of stupendous achievements in architecture, in public works, in dramatic arts, in ceramics, even in philosophy and logic. But in science proper, in physical science, no one had achieved, as already said, anything comparable to the achievements of our Western Christian Civilization.

The characteristic note of modern western science is its impressive capability to describe quantitatively a panoramic variety[4] of natural phenomena, from the realm of the elementary particles, to the evolution with time of the entire universe, through mathematical relations and differential equations, in a most precise and effective fashion. Nothing comparable had taken place in any of the great historical civilization, now extinct or alive. In our contemporary Western civilization, with roots in medieval Christendom generations of scientists have made scientific progress following in their work a middle road between too much empiricism and too much idealism. Regardless of their ideological bent, the creative Western physicists, chemists, cosmologists of the last three centuries, when doing their work, did it as realists. Only in a realist perspective is it possible to escape the pitfalls which early scientific developments in other civilizations found in their way.

In those civilizations, the universe was either chaotic and incomprehensible or subject to iron deterministic laws of eternal returns. In either care there was no point in dedicating strenuous and sustained efforts, generation after generation, to understand it. In medieval Christian Europe, on the other hand, the world, created by God from nothing and in time, was basically good. It was well done, and man's intellect well made by God in his image and likeness, and well suited to perceive the unity, the truth and the beauty in the world.

If only matter is real and subject to purely chaotic behaviour or to strictly deterministic laws, regardless of what the observer does or does not, everything will go on in its own chaotic or deterministic course. No need therefore for science. If, on the contrary, only ideas count, and the material world is radically false or perverse, there would be no point in the investigation of the natural laws which govern such a universe. The adventure of modern science has required indeed the efforts of many common men, and a handful of geniuses, through many generations. And it has been possible only through faith in the objective value of a universe well done, and confidence in man's intellectual ability to understand it.

If the world had not been rationally made, in other words, if something rationally valid one day were not valid next day, or if a thing rationally valid here were invalid in any other more or less close neighbourhood, systematic knowledge would be impossible.

It does not require to be a scientist to see the truth of these general considerations. The astonishing edifice of contemporary science, however, is a monumental proof of the fact that it is possible to describe quantitatively and systematically the laws governing nature, from the elementary particles to the most distant galaxies, and should be, therefore, a clear pointer to the wisdom of the Creator of nature and of the laws which govern it, as well as to the fact that man is endowed with intelligence and freedom. A freedom which allows him to recognize his Creator or to reject Him freely.

There is a subtle connection between a general theory of the physical world, a theory of knowledge and a natural theology, all three grounded in a Christian realism, heir of a revealed biblical wisdom which had come from the East and had infused a new life in the Western Greco-Roman Ecoumene at a time when the Roman Empire began to show signs of decline.

14.2. Early Medieval "Natural Philosophers"

Abelard of Bath (fl. 1125) was one of those eager early medievals who went on long and arduous journeys in the quest of learning. He travelled as far as Aleppo, in the Middle East. Through him medieval Europe got access to trigonometry, to the description of astrolabs and to Euclid's geometry. His contacts with Muslim learned men made him aware of the struggle in the Arab culture to reconcile faith and reason. Having in mind his acquaintances with medieval Muslim and Jewish learned men, Abelard is on record remarking to his nefew that many of his contemporaries identified God with Nature. At his time, which was intellectually a turbulent time (but not only at his time) there was a strong tendency in men to identify nature with the ultimate entity, a tendency to be overcome only if men were willing recognize humbly their dependence on a truly transcendent Creator. It is characteristic of Abelard in his "Questiones naturale"[5] the affirmation of the prerogatives of the Creation as well as those of the Creator. He favoured natural explanations over miraculous ones, whenever possible, showing his proclivity to a true scientific attitude facing the physical world: "I do not detract from God … Whatever there is, is from Him and through Him. But the realm of being is not a confused one … Only when reason totally fails, should the explanation of the matter be referred (directly) to God".

For medieval men as for men at anytime, the experience of personal freedom goes hand in hand with the experience of innate laziness. They too was tempted by the mirage of fatalism, which lured him towards astrological pantheism.

The biblical account of Creation, played a purifying role in the writings of Thierry of Chartres (d.c. 1155) rising him well above Greek animism and pantheism, even in its best literary expressions, such as Plato's "Timeus". "The intention of Moses (says Thierry) was to show that the creation of things and the formation of men was by the only one God, to whom alone worship is due. The utility of the work [of Moses] is the acquisition of knowledge about God through His handiwork". He saw Aristotle's four causes in a new perspective:

"Of the substances composing the world there are four causes: efficient, or God; formal, or God's wisdom; final, or His kindness; material, the four elements .." Note[6] that for Aristotle, as for the medievals, "Earth" included all solids, "water" meant all liquids, "air" all gases, and "fire" all heat. According to him all "elements", composing the world, "should have Him as their Creator, because they are all subject to change and they perish". For Thierry of Chartres, the speed of the moving heavens was much like the flight of a projectile: "when a stone is thrown, its impetus is ultimately due to the hold of the thrower against something solid [the ground]; the more firm is the stand of the one who throws, the more impetuous is the throw". For Thierry "there are four kinds of reasons that lead man to the recognition of his Creator: the proofs are taken from arithmetic, music [harmony], geometry and astronomy". If the Creator had actually arranged everything "according to number, measure, and weight", as recorded in the Book of Wisdom, man's understanding of the world had to reflect a mathematical character.

Robert Grosseteste (c. 1168–1253), Bishop of Lincoln, and possibly the first chancellor of the University of Oxford, a most influential figure in medieval thought, was even more explicit about the mathematical character of man's understanding of nature:
"The usefulness of considering lines, angles and figures is the greatest, because it is impossible to understand natural philosophy without them. They are efficacious throughout the universe as a whole and its parts, and in related properties as in rectilinear and linear motion". For instance, Grosseteste's investigation of the rainbow gives an excellent example of his scientific methodology, including seminal programs for induction, falsification, and verification. He, rightly, attributed the rainbows phenomenon to the refraction of light, rather than to its reflection, as done by Aristotle and Seneca. He considered all measurements made by man as imperfect, based as they were on "conventional units" and uncapable of counting the infinitely small, f.i. the (infinitely small) points contained in a line:
"... since these are infinite, therefore their number cannot be known by a creature but by God alone, who disposes everything in number, weight and measure". According to S.L. Jaki, his still unedited treatises show that Grosseteste's scientific methodology depended on the idea of the Creator as wholly rational, personal Planner, Builder, and Maintainer of the Universe.

Not all the writings of the natural philosophers of the thirteen century were, of course, as clear and rational as Grosseteste's. Many were a mixture of insights, rumours, sound principles and fantastic tales, a combination of reason and magic. But even in some of the most confusing cases one can perceive the miracle of an age which "did not wholly succumb to the magical and irrational" in the quest for understanding, "unlike all previous cultures". For instance, in "De universo" of William of Auvergne (d.c. 1249), bishop of Paris, one can find frequent references to magical and astrological aphorisms, but can find also a continued polemics "against Manicheism, fatalism, pantheism, star worship" and similar betrayals of man's rationality and ultimate destiny, identified by him with the Saracens (Arabs), the Greeks of old and the Hermetic philosophers.

The lengthy discussion in "De universo" of the Great Year, identified with the 26,000 years period completing the precession of the equinoxes, shows its author awareness of the fact that belief of "eternal returns" every Great Year represented for William the farthest possible embodiment of the pagan, non-Christian world view.

Defending right reason William, aiming at the contemporary Mutakallimun, those devout learned Muslim philosophers, says: "Similarly, you must also part with those who in such matters take refuge in almighty God's most imperious will, and wholly abandon these questions as insoluble, and feel themselves at ease when they say that the Creator willed it that way, or that His will is the sole cause of all such things", and he points to the fatal error of not distinguishing between primary and secondary causality.

Bibliography

1. S. L. Jaki, "The Origin of Science and the Science of its Origin" (Reguery/Gateway, Inc.: South Bend, Indian, 1978) pp. 22–42.
2. Fr. Louis Lecompte, "Nouveaux Memoires sur l'etat present de la Chine" (Chez Jean Anisson: Paris, 1696). [Quoted in Ref. 1]
3. S. L. Jaki, "Science and Creation: From Eternal Cycles to an Oscillating Universe" (University Press of America: Lanhan MD, 1990).
4. Julio A. Gonzalo, "Pioneros de la ciencia" (Palabra: Madrid, 2000).
5. S. L. Jaki, Ibid (Ref. 3) p. 219.
6. Virginia Trimble, FHP Newsletter, APS, Vol. X, No. 1, Fall 2006 p. 6.

Chapter 15

Tomas Aquinas and
Roger Bacon

15.1. Tomas Aquinas and the Ways to God

About the same time the great Saint Thomas Aquinas (1225–74) was embarked in a gigantic effort to bring together faith and reason "in a stable synthesis". Moderate realism characterized his theory of knowledge, and the analogy of being formed the essence of his metaphysics. His resolute commitment to give reason his due meant as much as possible generous acceptance of the Aristotelian system, for two millennia the epitome of a rational explanation of the world. This commitment was motivated, as noted by Stanley Jaki, by the contemporary predicament of Muslim theologians and philosophers.

Aquinas completed his "Summa contra gentiles" (1257) with the aim of countering the occasionalism and the fatalism at that time contending with one another within Muslim contemporary culture. The work centered on questions about the Creator and about the nature of human intellect, trying always to depart little from Aristotle, except when the Christian Creed allowed no room for an agreement.

In his "Summa Theologica" (1273), a work of synthesis, not polemics, Aquinas went to a surprising extent in accepting Aristotle's cosmology and physics. Even to counteract the idea of an infinite endurance for the world, he resorted only to theological arguments. For Aquinas, the ultimate "raison dêtre" of the cosmos consisted in its subordination to man's supernatural destiny. Contrary to Aquinas, his master, Albertus Magnus was an enthusiastic advocate of experimental investigation, who found in the contingency of the world the justification of his prolific work on natural history. There was no difference, of course, between the disciple and the master, on the all important doctrine rejecting the inevitability of "eternal recurrences" in the world. But Albertus was more explicit in his dissertation on "De fato", about the history of the question of "eternal returns" in Plato, Aristotle, the Stoics, Ptolemy, the Arabs (especially Albumasar) partisans all of them of the "eternal returns", as well as in the early Church Fathers, who, of course, opposed it urgently.

In his "Summa Theologica" Saint Thomas raised the question: Whether God exists? Then he first presents two momentous objections. Objection 1: the existence of evil; Objection 2: the explanation of everything by natural causes, which makes God superfluous. Then he proceeds with his famous Five Ways,[1] beginning, all of them with specific characteristics of actual, existing beings.

First:	From motion (change: generation/corruption).
	Arriving to a Prime Mover
Second:	From causality.
	Arriving to an Uncaused Cause.
Third:	From contingency.
	Arriving to a Necessary Being.
Fourth:	From the degrees of perfection.
	Arriving to a Most Perfect Being.
Fifth:	From order.
	Arriving to an Ordering Intellect.

As Saint Thomas says, this Prime Mover, Uncaused Cause, Necessary Being, Most Perfect Being, Ordering Intellect is what everyone understands to be God.

The most elegant and most metaphysical of the five is probably the Third Way: We find in nature contingent things (things which are possible to be or not to be). But it is impossible for these things always to exists, for that which is contingent cannot exist forever. Therefore, if everything were contingent, then, at one time there would be nothing in existence, because everything that exists begun to exist by something already existing. Which is absurd. Therefore the chain of contingent beings points necessary to a Necessary Being (Neither an infinite Chain made up only of contingent beings nor a closed circle of contingent beings explains the actual observed existence of contingent beings).

Take a star, for example, our Sun, a second or third generation star. We know today, Saint Thomas did not know (because at his time there were no NASA's artificial satellites, no Hubble's Telescopes, etc.), that an average mass star has a finite lifetime, perhaps ten thousand million years (longer if it is less massive, larger if it is more massive). This time is the time needed to consume all the hydrogen in its interior to produce helium and heavier elements. Then it will go through a phase as red giant, followed by another phase as white dwarf, to end up as a residual black dwarf. In case of stars more massive than the sun the resulting cosmic dust will become latter a newborn star, if physical conditions are propitious, under the gravitational attraction. And so forth... but not infinitely. The universe will be becoming more diluted and colder with the pass of time in a scale of tens of thousands of years. Stars are then contingent beings. And there is a finite number of generations of stars going back in time towards the Big Bang. We can trace back the matter and the energy to the time when atoms were first formed, out of the primeval plasma of electrons, protons and alpha particles. Then we can proceed backwards to the time when the elementary particles were formed in the presence of a very high radiation density. At these times no galaxies, no stars. And then we will approach times closer and closer to the Big Bang, till the uncertainty principle does not allow us to follow the time evolution of energy and material particles precisely.

The universe, therefore, is made up of contingent components, under the laws of physics (gravitation, radiation pressure, electromagnetic, strong and weak interactions ...) In all appearance the constituents of the universe as described by contemporary physics look contingent. The conservation of matter and energy (in ever more diluted form) the laws of physics, and the universal constants, all attest the cosmic stability and rationality. But Saint Thomas still is right: the actual existence of contingent beings requires the existence of a Necessary Being. A Necessary Being who is also, to begin with, a Prime Mover, an Uncaused Cause, a Most Perfect Being and an Ordering Intellect: "The wisdom of God (is) manifested in the works of Creation" as confirmed by the Book of Nature and the Book of Revelation.

15.2. Roger Bacon and the Experimental Method

Another influential man of the thirteen century was the franciscan Roger Bacon (1214–1294), medieval pioneer of the experimental method in the study of nature. Bacon's impetuous efforts to secure the service of science for the Christian faith lead him to compose his "Opus majus" (1267) which resulted in his temporary imprisionement, because some of his novel views stressing the inexorable determinism of events in nature appeared to his superiors as suspicious of incompatibility with man's freedom and moral responsibility. As noted by Stanley Jaki[1] his case shows clearly, nevertheless, "the difference between the Capitulation of most Arab commentators of Aristotle to the idea of cyclic determinism and Bacon's refusal to compromise in this crucial issue". He warned about the difference between final and efficient (secondary) causes, between man's supernatural destiny and his temporal (every day) subsistence. He asserted the forever partial character of man's knowledge about the world, in contrast to Aristotle's strict claims about *a priori*, definitive truths concerning the processes of nature.

In his "Opus majus" Bacon went as far as including sections on the application of astrology to Church government. Also, he was most emphatic on the existence of cycles in human history, paralleling the recurrence of planetary close encounters. But in the end he rejected the possibility of "eternal returns" of the world, because he saw them as incompatible with a Creator doing freely his work of creation. Taking the precession of the celestial axis (the period which he put as twenty-six thousand years) for evidence of the Great Year, would have been like following in the footsteps of Siger of Brabant (fl. 1260–77). This would have been going too far in the advocacy of recurring the autonomy of philosophy from theology, as done by Averroes, the most distinguished Arab commentator of Aristotle. For Averroes the world was, as it was for Aristotle, a necessary emanation of the First Cause. In his grandiose, pantheistic, organismic universe of things, equivalent to a supreme animal, every event was believed to come back again and again with inexorable circularity. Aquinas, who was not a polemicist by nature, came forward immediately to discredit the "murmurantes" (Siger and his followers) in a public lecture on the subject of the World's eternity.

Bibliogaphy

1. Peter Kreeft, "Summa of the Summa" (Ignatius Press: San Francisco, 1990).
2. S. L. Jaki, "Science and Creation: From Eternal Cycles to an Oscillating Universe" (University Press of America: Lanhan MD, 1990) p. 227.

Chapter 16

From Buridan and Oresme to Copernicus and Newton

In 1277 Etienne Tempier, bishop of Paris condemned 219 propositions covering a wide spectrum of questions clearly excluding the eternity of the world (Prop. 83–91) and the perennial recurrence of everything every 26,000 years (Prop. 92). In substance the bishop of Paris was affirming rigorously the contingency of the world with respect to a transcendental Creator, source of all rationality in the macrocosm as well as in the microcosm, in heaven as well as in Earth.

The decree bishop Tempier was taken by Pierre Duhem, the great French physicist and pioneer historian of physics, as the starting point of a new era in scientific thinking.

In the decree it was recognized the possibility of several worlds (something which perhaps could attract some sympathy from some present day partisans of extra-terrestrial intelligent beings) (Prop. 27); the rejection of superlunary bodies as animated, incorruptible and eternal (Prop. 31–32); the possibility of rectilinear motion for celestial bodies as part of a celestial machinery (Prop. 75); the rejection of a deterministic influence of celestial stars on the life of individual men from the instant of birth (Prop. 105); the rejection of the provenance of the "first matter" from celestial matter (Prop. 107); all of it aimed at defending the exclusive rights and freedom of the Creator when he created. This attitude shaped the conceptual framework for a revolutionary new approach towards the understanding of the motion of celestial as well as terrestrial bodies. The statement of Paris' bishop was not binding on the universal Church, but it was primarily a university affair, brought up by the fact that Siger of Brabante had been teaching for over a decade at the

University of Paris, at which bishop Tempier was charged with the duty of maintaining orthodoxy.

In his "Lowell Lectures" (1925) A. N. Whitehead[1] said: "I do not think however, that I have even yet brought up the greatest contribution to the formation of the scientific movement. I mean the inexpugnable belief that every detailed occurrence can be correlated with its antecedents in a perfectly definite manner, exemplifying general principles (emphasis added). Without this belief the incredible labours of scientists would be without hope. It is this instructive conviction, vividly posed before the imagination, which is the motive power of research: that there is a secret, a secret which can be unveiled. How has this conviction been so vividly implanted in the European mind?"

There was an intimate connection between the rational conjectures of medieval "physicists" and their firm believes in a personal Creator. This can be seen in their comments on "De Caelo" (On the Heavens), the systematic exposition of Aristotle's world views. Of extraordinary importance in this respect is John Buridan's "Questioness super quattor libris de caelo et mundo"[2] through its slightly modified version by his disciple Albert of Saxony, which, according to Stanley Jaki had year after an unmistakable influence on Galileo. Copies of this work were commonly available at medieval European universities, from Salamanca to Krakow.

The Aristotelian clear-cut distinction between superlunary and sublunary matter was dealt a decisive blow by Buridan. While Aristotle denied that the heaven could decay, Buridan reminded his readers that they could decay and even that the Creator was perfectly capable of annihilating the world.[3]

Buridan warned against confusing faith and reason, and offered an explanation of the plurality observed in the heavenly motions which did justice to both theology and natural science.

He noted that "in natural philosophy one should consider process and causal relationships as if they always come about in a natural fashion; God is no less the cause, therefore, if this world and its order have an end than if this world was eternal".

Actually he did not yet accept as a fact the rotation of the Earth on its axis (his discipline Oresme would do it latter) but, against Aristotle, he maintained the continuous though relatively small changes of the Earth respect the fixed stars, which, for him, implied that the Earth was not at the center of the universe. This and other considerations represented an enormous step in the right direction, away from the ancient pagan worldview epitomized by Aristotle.

Buridan, against the mistaken notions on motion as due to continuous contact of the mover with the moved body, and on the idea of the natural attraction of a body to its "natural" place, proposed his notion of "impetus", a quality implanted in the moving body by its mover, and the notion of "gravity", a property innate to all massive bodies. And, after reviewing the usefulness of his theory with respect to various movements down here on Earth, he further dared to outline its usefulness for celestial mechanics. In the same breath he wrote of jump, of planetary motion, and of the Creator as the ultimate agent imparting a given quantity of motion to the various parts of the universe:

"[If one] Whishes to jump a long distance [goes] back a way in order to run faster, so that... he might acquire an impetus which would carry him a longer distance in the jump... the person so running and jumping does not feel the air [as] moving him, but [rather] feels the air in front strongly resisting him.

Also, since the Bible does not say that appropriate intelligences move the celestial bodies, it could be said that it does no appear necessary to posit intelligences of this kind... [rather it could be said that] God, when He created the world, moved each of the celestial orbs as He pleased, and in moving them He impressed in them impetuses which moved them without His having to move them anymore except by the... general influence whereby He concurs as a co-agent in all things which take place... And these impetuses which He impressed in the celestial bodies were not decreased not corrupted afterwards, because there was no inclination [to it]... Nor was there resistance which would be corruptive or resistive of that impetus... But His I do not say assertively, but [rather tentatively]..."[4]

Nicole Oresme (1323–1382) was the most outstanding disciple of Burindan. At the request of Charles the V, king of France, Oresme undertook the translation and interpretation of the three major works of Aristotle, the "Nicomean Ethics", the "Politics" and "On the Heaven". While he was at the College of Navarre, of which he became grand master in 1536, he did very original work on a wide range of topics including monetary theory, astronomy, geometry and algebra. In his "De configurationibus qualitatum et motuum" he was trying to get a universal mathematical method to describe physical changes as wall as changes in man's inner experiences, psychological and esthetic. His commentary to "On the Heaven" is considered today as a great classic of scientific literature. Oresme's work has received encomiastic support from great pioneer historians of science such as P. Duhem and H. Dingler, but has received a more critical appraisal from other historians such as A. Koyré, one of the most influential positivist historians of science in the last century. Of course there is good justification in recognizing the newness of Galilean Science, but the primitive steps in the same direction in the writings of Burindan and Oresme deserve as much justified recognition, as a fact of history, with an enormous impact on the modern mind, proud of its science.

With reference to Aristotle's insistence on the perfection of the universe, Oresme said that the perfection of the laws of nature were but a modest reflection of the infinitely perfect attributes of the Creator. Oresme allowed incorruptibility to the celestial bodies and the celestial motions only in the restricted sense that these motions were frictionless, allowing a unified perspective to discuss earthly and heavenly motions. Against the possibility of eternal recurrences he noted in his "De proportionibus proportionum" that the periods of the planets are most likely incommensurable and, therefore, that they cannot return exactly to the same relative position. In any case, he noted, such periods should be much larger than the 26,000 years of the period of precession of the equinoxes, the length assigned in ancient Greece to the "Great Year" after which everything would be repeated again. Belief in the "Great Year" had yielded to a new era of thinking, rooted in the liberating influence of the doctrine about creation by an absolutely sovereign Creator. Oresme even speculated on the possibility of worlds enclosed within worlds. Oresme parted with Aristotelian necessitarianism implying that several words would require several Gods. He responded: "One God would govern all such worlds"[5]

According to Aristotle the Prime Mover's perfection imposed a strictly spherical shape on the world. Oresme considered this an unreasonable restriction on the Creators power, admitting the possibility of empty space either outside or inside the stellar realm. It was a very remarkable anticipation of the section in Newton's Scholium to his "Principia" in which the idea of an infinite empty space is conceptually linked to the infinity of the Creator. Oresme said: "...Outside the heavens, there is an empty incorporeal space quite different from any other plenum or corporeal space, just as the extent of this time called eternity is of a different sort than temporal duration, even if the latter were perpetual... Now this space of which we are talking is infinite and indivisible, and is the immensity of God and God himself. Just as the duration of God called eternity is infinite, indivisible, and God himself, as already stated above"[6]

Clearly, for Oresme, God was transcendent to the world, and his transcendence could be safeguarded by viewing him as imparting a given quantity of motion (impetus) to the world once and for all.

Later "impetus" would be correctly redefined as "momentum", i.e. inertial mass times the velocity imparted to the moving body. This crucial conceptual development was impossible to conceive within the pantheistic necessitarianism of Aristotle.

But the full process from Buridam and Oresme to Newton took three hundred years. Copernicus (1473–1543) with the impetus theory behind him, gave the crucial step of postulating the heliocentric description of the inertial planetary movements, with references to how the Psalmist said properly that God was immensely pleased by his handywork. Then Galileo (1564–1642) championed vigorously the Copernican system, and dis so nothing that according to the Church Fathers (Saint Augustine in particular) the Bible said "how to go to Heaven, not how the Heavens go". It may be pointed out also that his arguments based on the tides to support the Copernican system were flawed and that two hundred years would pass before the first evidence of the parallax of a star was observationally confirmed.

Fig. 16.1. Paris' Cathedral.

And Johannes Kepler (1571–1630), luckily named assistant to Tycho Brahe in Prague's observatory inherited Tycho's precise observational records of the planetary evolutions, leading to the formulation of his three famous laws: 1st) The planetary orbits are ellipses; 2nd) The areas swept by the vector radii in equal times are equal; and 3rd) The ratio between the cubes of the semi-major axis of the ellipses and their corresponding squared period were the same for all solar planets then known.

Everything was ready for Isaac Newton (1642–1727) to formulate precisely the laws of classical mechanics and the law of universal gravitation noting the analogy between the fall of an apple and the slow fall of the moon on Earth.

Thus, the centuries of medieval European Christendom, which witnessed the creation of the medieval universities: Bologna, Paris, Oxford, Salamanca, Prague, Krakow..., and the impressive gothic cathedrals: Chartres, Paris, Oviedo, Leon, Burgos...with their unsurpassed elegance and beauty, were also the centuries in which the roots of the edifice of modern science begun to take hold in the conceptual developments adumbrated by Buridam, Oresme and their medieval predecessors.

Bibliography

1. A. N. Whitehead, "Science and the Modern World: Lowell Lectures, 1925" (The Macmillan Company: New York, 1925).
2. Edited by E. A. Moody (Cambridge, Mass: The Medieval Academy of America, 1942).
3. Ibid., p. 90 (Lib. I, quaest. 14).
4. M. Clagett, "The Science of Mechanics in the Middle Ages" (University of Wisconsin Press: Madison, 1959) p. 536.
5. N. Oresme, "Le Livre du ciel et du monde", p. 171 (Book I, Chap. xxiv).
6. Ibid., p. 177.

Chapter 17

The Wisdom of God Manisfested in the Works of Creation

Robert Boyle (1627–1691), a contemporary of Newton, perhaps the first chemist in history, and discoverer of the law of the perfect gasses (the law of Boyle–Mariotte), is the author of a great book with this expressive title. "The wisdom of God manifested in the works of Creation".

Fig. 17.1. Boyle.

After Boyle and Newton, along the eighteenth and nineteenth century, the principles of what we know as modern science were developed by a long list of outstanding scientist. The divisions among Christian nations in Europe, the winds of the new age of the "Enlightenment", the French Revolution, and the subsequent Industrial Revolution, contributed to weaken the faith in Europe, and waves of anticlericalism swept the continent. It is worth nothing however that at that time (and later) the popular appeal of the Christian faith in America (English, Spanish and Portuguese speaking) did survive much better, and no such a thing as a widespread anti-clericalism was really dominant at any time. But all along these centuries most outstanding creative physicist, chemists, mathematicians, geologists, astronomers and biologists believed a personal God, whose wisdom was for them amply and unmistakenly manifested in the works of his Creation.

Any half-complete list doing justice to so many outstanding God believing scientists (catholic, protestant, Anglican, orthodox) in those centuries would be too long. Let us however brig up a representative sample of testimonies by a few truly great scientist, mostly European, same American, who manifest eloquently their faint in the Creator.

17.1. Physicists

Isaac Newton[1] (1642–1727): "This most beautiful system of the sun, planets and comets could only proceed from the counsel and dominion of an intelligent and powerful Being" ("Principia", 1687).

Alexander Volta[2] (1745–1827): "I do not understand how anyone can doubt the sincerity and constancy of my attachment to the religion which I profess, the Roman, Catholic and Apostolic religion in which I was born and brought up... I have, indeed, and only too often, failed in the performance of those good works that are the mark of a Catholic Christian, and I have been guilty of many sins: but, through the special mercy of God I have never, as far as I know, wavered in my faith... and I constantly give thanks to God, Who has infused into me this belief in which I desire to live and die, with the firm hope of eternal life." (Milan, Jan. 6, 1815).

André Marie Ampère[3] (1775–1836): "We can see only the works of the Creator but through them we rise to the knowledge of the Creator himself. Just as the real movements of the stars are hidden by their apparent movements, and yet it is by observation of the one that we determine the other: so God is in some sort hidden by His works, and yet it is through them that we discern Him and catch a hint of the Divine attributes".

Fig. 17.2. A. Volta and A. M. Ampère.

Michael Faraday[4] (1791–1867): "There is no philosophy in my religion. I am of a very small and despised sect of Christians… and our hope is founded on the faith that is in Christ… But, though the natural works of God can never by any possibility come in contradiction with the higher things that belong to our future existence… I do not think it at all necessary to tie the study of natural sciences and religion together, … (and) that which is religious and that which is philosophical have ever (for me) two distinct things" (Letters, October 24th, 1844).

James Clerk Maxwell[5] (1831–1879): "I have looked into most philosophical systems and I have seen that none will work without a God". And another of his saying during those his last days: "Old chap! I have read up many queer religions: there is nothing like the old think after all".

We may note that all four greatest contributors to the science of Electromagnetism, from its inception Volta, Ampere, Faraday and Maxwell, were God believing men and devote Christians, like Galvani, Coulomb, Ohm, Oersted and many others. And we do not live now in the "atomic" era but in the era of "electricity" and "electronics". Everyday life depends on large electric power stations, communications depend on electromagnetic waves, and even compact personal computers are fed by batteries, which can be traced back to Volta's piles. The units used to quantify and operate electric and electronic devices are "volts", "amperes", "farads", etc, in honor of those great Christian scientists.

Julius Robert Mayer[6] (1814–1878): "… that in the living brain, physical changes designated by the name of molecular activity are continually taking place, and that the spiritual functions of the individuals are connected in the most intimate manner with these cerebral processes. But it is a great blunder to identify these concurrent activities. An example will demonstrate it unequivocally. Telegraphic communication cannot, as every one knows, be established without a simultaneous chemical process. But the message delivered by the wire, the contents of the telegram, can by no means be regarded as a function of this electro-chemical process. This holds with still greater force of the relation of thought to the brain. The brain is not the soul, but only the instrument of the soul… not an object of investigation for Physics or Anatomy. Were it not for this inalterable harmony, pre-established by God, between subject and object, all our thinking would be necessarily without fruit. Logic is the Statics, Grammar the Mechanics and Language the Dynamics of thought."

Mayer is credited with the discovery of the principle of conservation of energy, and, with Joule, Holtzmann and Helmholtz, independently, of the mechanical equivalent of heat. In his "Kleinere Shrifen" on Dec. 31st, 1851, he had written: "My early feeling that scientific truths are to the Christian religion much what brooks and rivers are to the ocean, has become my most vital conviction."

Fig. 17.3. James C. Maxwell and Robert Mayer.

James Prescott Joule[7] (1818–1889): "I shall lose no time in repeating and extending these experiments, being satisfied that the grand agents of nature are, by the Creator's Fiat, indestructible, and that whatever mechanical force is expended, an exact equivalent of heat is always obtained" And, on the ignitions of meteorites upon entry into the atmosphere: "I cannot but be filled with admiration and gratitude for the wonderful provision made by the Author of Nature for the protection of his creatures. Were it not for the atmosphere which covers us with a shield, impenetrable in proportion to the violence which it is called upon to resist, we should be continually exposed to a bombardment of the most fatal and irresistible character... Even dust flying at such a velocity (18 miles per second) would kill any animal exposed to it".[8]

William Thomson (Lord Kelvin)[9] (1824–1907): "We must regard the sun as the source from which the mechanical energy of all the motions and heat of living creatures, and of the motions, heat and light derived from fires and artificial flames is supplied"... "All modes of life derive their bodily heat from chemical changes which are due to the food which they have taken. The food from man and from animals comes ultimately from the vegetable kingdom, that of the herbivorous creatures directly, and that of the carnivorous indirectly. But how do plants grow and build up their organisms? Except the fungus, they grow in the greater part of their surface from the air and the soil by the decomposition of carbonic acid and water. This decomposition can however only go under the influence of light. Thus the plant is directly the child of the sun, as is the plant eating animal. Further more, form the sun is derived every species of artificial light and heat". "Coal composed as it is of the relics of ancient vegetation, derived its potential energy from the light of distant ages. Wood fires give us heat and light which has been got from the sun a few years ago. Our coalfires and gas lamps bring us for our present comfort heat and light of a primeval sun which have lain dormant as potential energy beneath seas and mountains for countless ages". And in another place10 he says: "It is impossible to understand either the beginning or the continuance of life without an overruling creative power". "I need scarcely say that the beginning and maintenance of life on Earth is absolutely and infinitely beyond the range of all sound speculation in dynamical science. The only contribution of dynamics to theoretical biology is absolute negation of automatic commencement or automatic maintenance of life".

Fig. 17.4. James Prescott Joule and Lord Kelvin.

17.2. Chemists

Antonie L. Lavoisier[11] (1743–1794): "This great name of Lavoisier must be particularly remembered in our Society because the illustrious chemist was always a man of faith. This is the result of all documents on his life found in recent years".

Lavoisier is considered unanimously the founder of modern Chemistry. He was decapitated at the "guillotine" during the French Revolution.

John Dalton[12] (1766–1844): "In Dalton the character of the man equaled the superiority of his lights: he was a model of virtue without ostentation, of religion without fanatism."

He laid the foundation of atomic theory and contributed decisively to establish the laws of definite and multiple proportions for chemical reactions. He was very much appreciated by his contemporaries. When he died, more than 40,000 people came to pay the last tribute to his mortal remains. His disgust with the atheistic systems which were beginning to be popular at his time was evidenced in his great chemical treatise: "New System of Chemical Philosophy".

Jöns Jacob Berzelius[13] (1779–1848): "An incomprehensible force, foreing to those of dead matter, has introduced the principle (of life) into the inorganic world. This has been effected not by chance, but with the striking variety and supreme vision of a plan designed to produce definite results, and to maintain an unbroken succession of transient individuals which are born one from another, and which at their death bequeth their decomposed constituents to the formation of new organisms. Every process of organic nature proclaims a wise purpose, and bears the stamp of a guiding mind; and man, comparing the calculations he makes to attain certain ends with those he finds in organic nature has been lead to regard his faculty of thought and calculation as a reflection of the Being to whom he owes his existence. And yet it has happened more than once that a philosophy, proud all the while of its own profundity, has maintained that all this was the work of chance, and that such organisms only held their ground as had accidentally acquired the power of self-preservation and reproduction. But the advocates of such systems do not perceive that the element in nature which they call 'chance' is a thing physically impossible. Everything which exists springs from a cause, an operative force, this latter tending (like desire) to break into activity and secure for itself satisfaction... a process which in no degree corresponds to our idea of 'chance'. It will be more honorable for man to admire the wisdom which he cannot rival, than to puff himself up with philosophical arrogance and attempt with his paltry reasoning to penetrate mysteries which will probably remain for ever beyond the scope of human reason."

Berzelius was one of the early pioneers in chemistry and electrochemistry and he was responsible for the chemical nomenclature for elements and compounds based upon the Latin names of the elements.

17.3. Mathematicians

Karl Friedrich Gauss[14] (1777–1855): "There is in this world a joy of the intellect which finds satisfaction in science, and a joy of the heart which manifests itself in the aid men give one another against the troubles and trials of life. But for the Supreme Being to have created existences, and stationed them in various spheres in order to taste these joys for some 80 or 90 years- that were surely a miserable plan…" "Whether the soul were to life for 80 years or for 80 million years, if it were domed in the end to perish, such an existence would only be a respite. In the end it would drop out of being. We are thus impelled to the conclusion to which so many things point, although they do not amount to a coercive scientific proof, that besides this material world there exists another purely spiritual order of things, with activities as various as the present, and that this world of spirit we shall one day inherit". This divine wisdom (says a biographer) was the food and discipline of his soul up to that last silent midnight in which his eyes were closed for ever.

Gauss, called "the prince of mathematics", was one of the very few undisputable mathematical geniuses of all times.

Augustin Louis Cauchy[15] (1789–1857): "I am a Christian, that is to say, I believe in the divinity of Jesus Christ as did Tycho Brahe, Copernicus, Descartes, Newton, Fermat, Leibnitz, Pascal, Grimaldi, Euler, Guldin, Boscovich, Gerdil; as did all the great astronomers, physicist and geometricians of past ages: nay more, I am like the greater part of these a Catholic: and were I asked for the reasons of my faith I would willingly give them. I would show that my convictions have their source not in mere prejudice but in reason and resolute enquiry. I am a sincere Catholic as it were Corneille, Racine, La Bruyère, Bossnet, Bourdaloue, Fénelon, as were and still are so many of the most distinguished men of our time, so many of those who have done most of the honor of out science, philosophy and literature, and have conferred such brilliant lustre on our Academies. I share the deep conviction openly manifested in words, deeds and writings by so many savants of the first rank, by a Ruffini, a Haüy, a Laënnec, an Ampere, a Pelletier, a Freycinet, a Coriolis and I avoid naming any of those living, for fear of paining their modesty. I may at least be allowed to say that I loved to recognize all the noble generosity of the Christian Faith in my illustrious friends the creator of Crystallography (Haüy), the introducers of quinine and the stethoscope (Pelletier and Laënnec), the famous voyager on board of the 'Urania', and the immortal founders of the theory of Dynamic Electricity (Frencinet and Ampère)."

Fig. 17.5. K. F. Gauss and A. Cauchy.

According[16] to his biographer C. A. Valson, Cauchy was ever ready at the call of events "to defend or propagate religion or to set on foot works of charity". He was a zealous member of the Conference of Saint Vincent of Paul, and did all that he could to relieve the needs of others.

Bernard Riemann[17] (1826–1866): Another one of the greatest mathematicians of all time. With his contributions to non Euclidean geometry, lied down the mathematical groundwork for Einstein's Theory of Relativity. He died of tuberculosis not yet forty years old, and in the biographical sketch prefixed to his collected works it is registered the following: "His wife was saying to him the Our Father, he was no longer able to speak: but at the words "Forgive our trespasses", he raised his eyes aloft: and after two or three gaps, his noble heart had ceased to beat. The pious habits which he had learned in his childhood remained with him all his life, and he served God loyally, if not always after the orthodox forms. Daily self-examination in the presence of God he regarded... as one of the elements of religion."

17.4. Geologists and Geographers

George Cuvier[18] (1769–1832): "Our sacred books bring before our eyes at the very outset the Creator bidding his creatures pass under the gaze of the first man and commanding him to give them names." And discerning in it a deep allegorical meaning he adds: "It teaches us clearly that one of our first duties is to cultivate a sense of the goodness and wisdom of the Author of Nature by a continuous study of the works of His power."

Cuvier, a Protestant, believed in a first cause which directs the destinies of all creatures and has foreseen and ordered all. His pioneering work on fossils was collected in his famous "Treatise on Fossils."

Matthew Fontaine Maury[19] (1806–1876): "Astronomy is grand and sublime, but Astronomy overpowers with its infinities. Physical Geography charms with its wonders, and delights with the benignity of its economy. Astronomy ignores the existence of man; Physical Geography confesses that it is based on the Biblical doctrine that the Earth was made for man…"

Maury, native of Fredericksbourg, Virginia (USA), was the main promoter of an International Congress in Bruxels in which a universal record of transoceanic travel was established, which allowed the making of navigation charts about winds and sea streams which were very useful to shorten transatlantic travel and then transpacific and transoceanic travel in general. For him, a devout Protestant science and religion were so mutually helpful that neither one could go far without the other.[20]

James Dwigth Dana[21] (1813–1895): "The record in the Bible is therefore profoundly philosophical in the scheme of creation which it presents. It is both true and divine. It is a declaration of authorship, both of Creation and the Bible, on the first page of the sacred volume. There can be no real conflict between the two books of the Great Author. Both are revelations made by Him to Man — the earlier telling of God made harmonies, coming up from the deep past and raised to their height when Man appeared; the latter teaching Man's relation to his Maker and speaking of loftier harmonies in the eternal future."

Dana, native of Utica, New York was, according to von Zittel,[22] "a most brilliant geologist, zoologist and mineralogist. His studies of the geology of the Pacific, the volcanoes, islands coral formations and species of zoophytes and crustacean, stand in the forefront of the geological literature."

17.5. Astronomers

Sir William Herschel[23] (1738–1822): "Half a dozen experiments made with judgment by a person who reasons well are worth a thousand random observation of insignificant matters of fact" [...] "(Those of us who love wisdom) by metaphysics are enabled to prove the existence of a first cause, the infinite author of all dependent beings."

He is considered the greatest astronomer of the seventeenth century. He discovered first that the law of gravitation holds outside our solar system by investigating binary stars in nearby galaxies.

Heinrich Wilhem Matthaüs Olbers[24] (1758–1840): "I am grateful for the easy circumstances with which Providence has blessed me, which permit me to spend my last years in otio cumdignitate... I (have) tasted and enjoyed... all good that this earthy life has to offer and (am) now able to take (my) departure without reluctance" (From a letter in his old age addressed to his friend Frederick W. Bessel).

Olbers was the first to point out the paradox of the dark night as incompatible with an infinite universe.

Frederick William Bessel[25] (1784–1846): "I am learning to know more and more that the true darlings of fortune are those whom Heaven has blessed with a friend who accepts that name as something more than a mere commonplace" (Letter to Olbers) And, in another occasion: "Let us enjoy what God, Whose goodness is so infinitely beyond that of man, has given us [...] God knows how hard it is for me to be so near to you and yet not able to visit you."

Bessel was, with Struve, one of the discoverers of the stellar parallax (in Lirae and 61 Cygny), a discovery which was qualified by J. F. W. Herschel, the son of Sir Willian Herschel, and himself another great astronomer, as "the greatest and most glorious triumph ever seen by practical astronomy."

****

As well documented by Pierre Duhem and, more recently, by Stanley Jaki, contemporary science (physics in particular) has unmistakable Christian roots.

Who were the true pioneers of this age of electricity? Coulomb, Volta, Ampere, Faraday, Maxwell great Christian scientists, all of them.

Also, the two greatest 20th century physicists, Planck and Einstein, in the best tradition of Western science, were objectivists and realists. Planck who was no friend of Mach's subjective positivism, held that the investigation of the laws capable of describing "absolute reality" always seemed to him "the most beautiful task for a scientist in this life".

Einstein on the other hand affirmed that "every two legged animal (every man) was in fact a metaphysicians". He said also that he would have preferred his "Theory of Relativity", to be known as the "Theory of Invariance" based as it was on the constancy of the speed of light.

Today, many excellent scientist are agnostic or atheist, but it is easy to check that from the last century onwards a good number of the distinguished physicist Nobel Prize Winners, Protestant, Catholic and Jews are on record as believing in God (the God of Abraham, Isaac and Jacob, Who is the God of Jesus Christ): J. J. Thomson (discoverer of the electron) The Bragg's (pioneers of X-ray diffraction), A. H. Compton (discoverer of the particle-wave duality of light quanta) P. Debye (pioneer of Quantum Mechanics), V. F. Hess (pioneer of cosmic ray research), Ch. Townes and T. H. Mainman (co-inventors of the laser), A. Penzias (co-discoverer of the cosmic background radiation) and B. Brockhouse (pioneer of neutron spectroscopy). To these, other Nobel Prize Winners in Chemistry, Biology and Medicine could be added.

Fig. 17.6. A. Penzias.

These men and many other are giving testimony of the deep harmony between Christian faith and contemporary science. Some of them are also on record in the defense of the sacred character of human life from conception to natural death.

Bibliography

1. Sir Isaac Newton, "Mathematical Principles of Natural Philosophy and His System of the World", translated and edited by F. Cajori, p. 544 (University of California Press: Berkeley 1934).
2. K. A. Kneller, "Christianity and the Leaders of Modern Science" with an Introductory Essay by Stanley L. Jaki (Real View Books: Fraser Michigan, 1995) p. 117.
3. Ibid., p. 123.
4. Ibid., p. 126.
5. Ibid., p. 136.
6. Ibid., p. 17.
7. Ibid., p. 23.
8. Ibid., p. 25.
9. Ibid., p. 32.
10. Ibid., p. 34.
11. Ibid., p. 180.
12. Ibid., p. 180.
13. Ibid., pp. 180–181.
14. Ibid., pp. 48–49.
15. Ibid., pp. 54–55.
16. Ibid., p. 50.
17. Ibid., pp. 61–62.
18. "Nature", 9, p. 340, August 1888.
19. K. A. Kneller, Ibid., p. 223.
20. Ibid., p. 276.
21. Ibid., p. 275.
22. "The Scientific Papers of Sir William Herschel", ed. J. L. Dreyer (Royal Society: London, 1912), 1: lxxxi.
23. K. A. Kneller, Ibid., p. 88.
24. Ibid., p. 89.
25. Ibid., p. 327.

Cosmic Numbers and Concluding Remarks

Chapter 18

Cosmic Numbers

As Max Planck[1] said in said in Stockholm (June 2, 1920) at the time he received the Nobel Prize for his discovery of the energy quanta: "Numbers decide".

The unprecedented accuracy of the cosmological data obtained by NASA's space satellites (first COBE, then Hubble, more recently WMAP) and by ESA's satellite Hipparcos have decided conclusively in favor of the Big Bang concept. It is possible now to describe cosmic events from very early times (baryon formation, electron formation, nucleosynthesis) to the all important event of atom formation (at which the universe transformed from hot plasma to a quasi-homogeneous atomic gas with minute thermal fluctuations). From there it is possible to proceed to the subsequent formation of stars and galaxies, and finally to the time at which the Sun and the Earth were formed.

The universe appears to be open (k < 0). Its total mass appears to be finite. It must have gone through an early "plasma" phase and is now in a relatively early stage of an atomic, transparent phase. Its mass density parameter must have been time-dependent $\Omega(t)$, as it must be also its Hubble parameter, $H(t)$.

But there are many other fundamental questions which, of course, remain open.

It is interesting to note that from the early 20th century onwards, physicists have tried more or less successfully to put the universe in numbers. A case in point is Sir Arthur Eddington,[2] who made early a "purely" mathematical estimate of the total number of baryons in the universe, giving it as the number

$$N_b = 2^{2^8} = 2^{256} = 1.157... \times 10^{77} \qquad (18.1)$$

which came out relatively close to

$$N_b = \frac{M_u}{m_p} \approx \frac{4.50 \times 10^{54}}{1.67 \times 10^{-24}} = 2.67 \times 10^{78} \qquad (18.2)$$

obtained using[3] $M_u = 4.50 \times 10^{54}$ g as the total mass of the universe, which is not far from the estimated number of galaxies (10^{11}), times the estimated number of stars per typical galaxy (10^{11}), times the average mass of a typical star like our Sun ($m_p = 1.67 \times 10^{-24}$ g is the baryon mass).

Fig. 18.1. A. Eddington.

Cosmic numbers evidently are determined to a large extent by the four interactions observed in nature. They can be evaluated quantitatively making use of the following modified version of Heisenberg uncertainty principle,

$$\Delta E \cdot \Delta t \approx \alpha \cdot \hbar \qquad (18.3)$$

Here the strength of each interaction is specified through the characteristic dimensionless constant (α) which appears multiplying Planck's constant ($\hbar$).

1. For the electromagnetic interaction:

$$\left(\frac{e^2}{r}\right)\cdot\left(\frac{r}{c}\right) \approx \alpha\cdot\hbar \rightarrow \alpha = \frac{e^2}{\hbar c} = 0.00731 \approx \frac{1}{137} \text{ (dimensionless)} \qquad (18.4)$$

2. For the nuclear strong interaction:

$$\left(g^2\frac{e^{-r/r'}}{r}\right)\cdot\left(\frac{r}{c}\right) \approx \alpha\cdot\hbar$$

$$\rightarrow \alpha_{NS} = \frac{g^2 e^{-r/r'}}{\hbar c} \approx 15e^{-2} \approx 2 \text{ (dimensionless)} \qquad (18.5)$$

3. For the nuclear weak interaction, responsible for the neutron decay (and indirectly for the primordial ^{4}He abundance) there is not available an expression for the potential energy.

4. For the gravitational interaction:

$$\left(\frac{Gm_p^2}{r}\right)\cdot\left(\frac{r}{c}\right) \approx \alpha_G\cdot\hbar \rightarrow \alpha_G = \frac{Gm_p^2}{\hbar c} = 5.9\times10^{-39} \text{ (dimensionless)} \quad (18.6)$$

There are some cosmic numbers, for instance the present cosmic "age" t_0 and cosmic expansion rate H_0 ($t_0 \approx [13.7 \pm 0.2] \times 10^9$ years), and the present photon to baryon ratio ($n_r(t > t_{af})/n_b(t > t_{af}) \sim 1.30 \times 10^9$), which are generally accepted and well established. Others, like the estimated physical conditions (cosmic time, cosmic density) at the time of ^{4}He nucleosynthesis, are generally accepted (with some what larger uncertainties) but they can be obtained assumming either a change or no change in equation of state at the time of atom formation, complementing them with either the value of the neutron half life or with the corresponding cosmic expansion rate. Finally others, like the widely assumed value for the cosmic density parameter $\Omega = 1$ for all time, distributed in the form of $\Omega_x = 1/3$ (dark matter) and $\Omega_\Lambda = 2/3$ (dark energy) are hope not yet incontrovertible.

What we can call the standard hot Big Bang approach to cosmic quantities is good enough yo account for the generally accepted values of t_0, H_0, (n_{r0}/n_{b0}), at present as well as for (ρ_r, ρ_m) at decoupling, $(\rho_r, \rho_m)_+$ at galaxy formation, etc, with a zero cosmological constant. It is good enough also to account for the cosmic time and cosmic density at nucleosynthesis, provided that a change in cosmic equation of state takes place at atom formation. (At times prior to decoupling/atom formation, the universe is assumed to be an expanding plasma made up of equal amounts of matter and energy). This standard approach does not allow for the time being for a satisfactory explanation of the substantial expected amount of missing mass and thee observed cosmic acceleration for relatively nearby galaxies with redshift $0.5 < z < 1.0$.

The standard hot Big Bang approach, complemented with the plasma equation of state, allows to speculate about earlier events such as electron formation and baryon formation. It is reasonable to point out that when cosmic density was much higher than the actual density of a collection of material particles put together (particles which are subject to exclusion principles, like Pauli's) it is not very meaningful talking about particles as, such or particle-antiparticle pairs. In principle it is meaningful to take only about non-diferenciated matter and non-diferenciated energy.

Immediately after the Big Bang, when cosmic temperatures as well as matter and energy densities were inconceivably high, the usual elementary particles (baryons, electrons ...) could not yet exit as such, because their characteristic mass density is much lower than the cosmic mass density at those very early times.

It is reasonable (and natural) to assume[4] (but far from generally accepted) that in this phase of the cosmic expansion, going through (i) baryon formation, (ii) electron formation and (iii) primordial nucleosynthesis, radiation energy density (ρ_r) and matter energy density (ρ_m) were equal to each other up to the time of "decoupling". As we have seen (Chapter 11) decoupling coincides with atom formation at a cosmic temperature of about $T_{af} = 2130$ K using the data available in

2003. The cosmic equation of state during this plasma phase would be given by

$$\rho_r(t) = \rho_r(t_{bf})(R_{bf}/R)^3 \text{ and } RT^{4/3} = \text{constant} \ (t_{bf} < t < t_{af}) \quad (18.7)$$

where the Stefan-Boltzman law ($\rho_r = \sigma T^4$) has been used. After decoupling (atom formation) the universe becomes transparent and radiation energy density (ρ_r) begins to decrease rapidly with respect to matter energy density. In this transparent phase of cosmic expansion, going through (iv) atom formation, (v) stars and galaxies formation and finally (vi) planet (Earth) formation, the cosmic equation of state is given by

$$\rho_r(t) = \rho_r(t_{af})(R_{af}/R)^4 \text{ and } RT = \text{constant} \ (t_{af} < t \le t_0) \quad (18.8)$$

All the way through the expansion, cosmic radius (R) and cosmic time (t) are related to each other by Einstein's cosmological equations

$$R = R_+ \sinh^2 y, t = (R_+/c|k|^{1/2})[\sinh y \cosh y - y] \quad (18.9)$$

where $R_+ = 2GM_u/c^2 = 6.08 \times 10^{26}$ cm. At early times, $y \ll 1$, therefore, $y \approx (R/R_+)^{1/2} \approx (T_+/T_{af})^{1/7}(T_{af}/R)^{2/3}$, using the "plasma" equation of state. Then

$$R \approx R_+(T_+/T_{af})(T_{af}/T)^{4/3} = R_{af}(T_{af}/T)^{4/3}, \ t \approx t_{af}(T_{af}/T)^2 \quad (18.10)$$

where[5] $R_{af} = R_+(T_+/T_{af}) = 1.72 \times 10^{25}$ cm is the cosmic radius at atom formation (decoupling) and $t_{af} = 9.62 \times 10^{13}$ s the corresponding cosmic time.

At all times and temperatures the baryon density, $n_b(t)$, is obtained dividing the mass matter energy density, $\rho_r(t)$, by $m_p c2$, where mp is the baryon mass, $m_p = 1.67 \times 10^{-24}$ g, and the photon density, $n_r(t)$, likewise, dividing the radiation energy density, $\rho_r(t)$, by ($2.8 k_B T$), which is the representative photon energy. We will check below the photon/baryon ratio at several important cosmic temperatures.

We begin evaluating tentatively some characteristic cosmic numbers. It will be seen among other things that some microscopic ratios such as the proton mass to electron mass ratio $m_p/m_e = 1836$, appears to be consistent with the cosmic dimensions at baryon $R(T_{bf})$ and electron formation $R(T_{ef})$, a reasonable result that one could not take for granted *a priori* unless the cosmological equation and the pertinent equations of state describe correctly true cosmic features.

(i) Baryon formation time (T = T_{bf} = $m_b c^2/2.8 k_B T$ = 3.88 × 10^{12} K)
Since the radiation (photon) energy density equals the matter (baryon) energy density, the ratio of photons to baryons is given by

$$\frac{n_r(t_{bf})}{n_b(t_{bf})} = \frac{\rho_r(t_{bf})/2.8k_B T_{bf}}{\rho_m(t_{bf})/m_b c^2} = 1 \tag{18.11}$$

Using r_b=1.24×10^{-13}cm for the baryon radius, we get for the total number of baryon formation and there after

$$N_b = \left(\frac{R_b}{r_b}\right)^3 = \frac{R_{af}^3 (T_{af}/T_{bf})^4}{r_b^3} = 2.67 \times 10^{78} \tag{18.12}$$

in excellent agreement with Eq. (18.1)

(ii) Electron formation time (T = T_{ef} = $m_e c^2/2.8 k_B$ = 2.11 × 10^9 K)
At this much lower temperature, the ratio of photons to baryons is much lower than one, but given that electrons are much lighter than baryons

$$\frac{n_r(t_{ef})}{n_b(t_{ef})} = \frac{\rho_r(t_{ef})/2.8k_B T_{ef}}{\rho_m(t_{ef})/m_b c^2} = \frac{m_b}{m_e} = 1836 \tag{18.13}$$

And, using r_e = 0.28 × 10^{-8}, of the order of Bohr radius which is the radius of the hydrogen atom (made up of electron and a tiny proton in the middle) the resulting total number of electrons in the universe is

$$N_e = \left(\frac{R_e}{r_e}\right)^3 = \frac{R_{af}^3 (T_{af}/T_{ef})^4}{r_e^3} = 2.67 \times 10^{78} \tag{18.14}$$

consistent with an overall cosmic charge neutrality (N_e = N_b).

(iii) Nucleosynthesis time ($T = T_{ns}$ = 4.6 × 10^8 K)
The cosmic time at nucleosynthesis[6] is tied to the free neutron half life τ_n = 887 s, and to the ^{4}He cosmic abundance consistent with a neutron/proton ratio of 0.131. According to fusion research data, the ignition temperature is of the order of T_{i0} = 39.8 KeV ≈ 4.60 × 10^8 K

and, which, using Eq. (18.10) allows obtaining the estimate of the time at nucleosynthesis as

$$t_{ns} = t_{af}(T_{af}/T)^2 = 2.06 \times 10^3 s \tag{18.15}$$

This leads to a neutron proton ratio

$$n/p = e^{t_{ns}/\tau_n} = 0.098 \pm 0.035 \tag{18.16}$$

consistent with the observed cosmic (primordial) ^{4}He abundance. This points to another dimensionless cosmic number

$$\frac{t_{ns}}{\tau} \approx 2.32... \tag{18.17}$$

Which is order one, with a considerable global uncertainty.

The next important steps in cosmic evolution are decoupling time and atom formation time. We note as very relevant the closeness of both times and temperatures, as pointed out in Chapter 11.

(iv) Decoupling and Atom Formation times ($T_{dec} = 2.135 \times 10^3$ K; $T_{af} = 2.968 \times 10^3$ K)
Using $H_0 = 71 \pm 4$ Km/s·Mpc, $t_0 = 13.7 \times 10^9$ years as reported in 2003 by WMAP, the temperature at which radiation and matter densities are equal (decoupling) is

$$T_{dec} = T_0\Omega_0\left(\frac{3H_0}{8\pi G}\right)\Big/\left(\frac{\sigma T_0^4}{c^2}\right) = 2135 \text{ K} \tag{18.18}$$

Using Saha's law the temperature at which there is 50% ionization/recombination in the cosmic plasma was given in "Acta Cosmologica" (1998) as $T_{af} = 3880$ K. WMAP's data on the other hand leads to $T_{af} = 2968$ K, not too far.

This time (decoupling, atom formation) is the latest time at which the ratio $\rho_r(t_{af})/\rho_m(t_{af}) = 1$ is still constant. Thereafter $\rho_r(t)/\rho_m(t)$ is less than one and tends to zero with increasing time.

On the other hand, this time (decoupling, atom formation) is the first time at which the photon to baryon ratio

$$\frac{n_r(t_{af})}{n_b(t_{af})} = \frac{\rho_r(t_{af})/2.8k_BT_{af}}{\rho_m(t_{af})/m_bc^2} = \frac{m_bc^2}{2.8k_BT_{af}} = \frac{T_{bf}}{T_{af}} = 1.30\times10^9 \qquad (18.19)$$

Becomes fixed and remains the same for $t > t_{af}$. This fixes also the total number of photons in the universe

$$N_r = \frac{n_r}{n_b}N_b = 3.47\times10^{87} \qquad (18.20)$$

(v) Galaxy formation time ($T_+ = 62.17$ K)
At $R_+ = 2\,GM_u/c^2 = 6.09\times10^{26}$ cm, large chunks of atomic gas become gravitationally bound for the first time in the expansion. Taking into account that $y = \sinh^{-1}(T_+/T)^{1/2}$ and that therefore $y_+ = \sinh^{-1}(1)$, we get for the corresponding time

$$t_+ = \frac{R_+}{c|k|^{1/2}}[\sinh y_+ \cosh y_+ - y_+] = 1.08\times10^{16}\text{ sec}$$
$$= 0.342\times10^9\text{ years} \qquad (18.21)$$

At this time the ratio of radiation energy density to matter energy density is already much less than one,

$$\frac{\rho_r(t_+)}{\rho_m(t_+)} = \frac{\rho_r(t_{af})(T_+/T_{af})^4}{\rho_m(t_{af})(T_+/T_{af})^3} = \frac{T_+}{T_{af}} = 0.0291 \qquad (18.22)$$

While the photon to baryon ratio remains

$$\frac{n_r(t_+)}{n_b(t_+)} = 1.30\times10^9 \qquad (18.23)$$

(vi) Present time ($T_0 = 2.726$ K)
At present (CBR temperature $T_0 = 2.726$ K as given by COBE) the cosmic radius and cosmic time are given by

$$R(t_0) = R_+ \sinh y_0 = R_+(T_+/T_0) = 1.389\times10^{28}\text{ cm} \qquad (18.24)$$

$$t_0 = \frac{R_+}{c|k|^{1/2}}\left[\sinh y_0 \cosh y_0 - y_0\right] = 4.27 \times 10^{17} \text{ sec}$$

(18.25)

$$= 13.7 \times 10^9 \text{ years}$$

where[7] $y_0 = 2.267$ has been used.
At this time

$$\frac{\rho_r(t_0)}{\rho_m(t_0)} = \frac{T_+}{T_0} = 0.00128$$

(18.26)

And the photon/baryon ratio remains

$$\frac{n_r(t_0)}{n_b(t_0)} = 1.30 \times 10^9$$

(18.27)

The accompanying Table 18.1 gives the characteristic mass, radius and time elapsed since the Big Bang at galaxy/star formation time in terms of natural Planck's units[8]

$$m_{pl} = \left(\hbar c / G\right)^{1/2} = 2.17 \times 10^{-5} \text{g}$$

$$l_{pl} = \left(\hbar G / c^3\right)^{12} = 1.61 \times 10^{-33} \text{cm}$$

(18.28)

$$t_{pl} = \frac{l_{pl}}{c} = 5.36 \times 10^{-44} \, s$$

Table 18.1. Cosmic quantities in natural Planck's units.

	Cosmic quantity	Natural unit	Ratio
Mass	$M_u = 4.11 \times 10^{54}$ g	$m_{pl} = 2.17 \times 10^{-5}$	1.89×10^{59}
Length	$R_+ = 6.09 \times 10^{26}$ cm	$l_{pl} = 1.61 \times 10^{-33}$	3.78×10^{59}
Time	$t_+ = 4.27 \times 10^{17}$ s	$t_{pl} = 5.36 \times 10^{-44}$	7.96×10^{60}

Most cosmic numbers evaluated above correspond to the early phase of expansion of the universe (beyond baryon formation) where the role of an hypotetic cosmological constant ($\Lambda \neq 0$) is generally assumed to be neglible. At those early times the hot standard Big Bang model is expected to give a good quantitative description of cosmic events.

Since 1998 observed galaxy redshifts in the interval $0.4 < z < 1.7$ support a cosmic accelerated expansion.[9] There are little data presently available from $z \approx 1$ to $z_+ = 21.8$, corresponding to the fist time of galaxy and star formation. For Martin Rees[10] the case for a non-zero cosmological constant is strong but not overwhelming. It may be noted that some arguments[11] against cosmic acceleration can be shown to be defective or inconclusive. But the history of up and downs of Λ through the years recommends caution[12] before ruling out alternative interpretations of the partial data available.

Bibliography

1. The English translation of Planck's speech in "Nobel Lectures. Including Presentation Speeches and Laureate Biographies – Physics 1901–1920 (Amsterdam: Elsevier, 1967)" Quoted by Stanley L. Jaki, "Numbers decide or Planck's constant and some constants of philosophy", in "Planck's constant: 1900-200" (Madrid: UAM, 2000), Julio A. Gonzalo ed.
2. E. T. Whittaker, "Eddinton's Theory of the Constants of Nature", the Mathematical Gazette 29, (1945), pp. 137–44. Quoted by Stanley L. Jaki in "God and the cosmologists" (Real View Books: Frase, Michigan, 1998) p. 103.
3. Julio A. Gonzalo, "Inflationary Cosmology Revisited" (World Scientific: Singapore, 2005) p. 78.
4. Ibid., pp. 69–72.
5. Ibid., pp. 79.
6. Ibid., pp. 75–76.
7. Ibid., p. 74.
8. Julio A. Gonzalo, "Planck's constant and the fundamental physical constants" in "Planck's constant 1900–2000" (Madrid: UAM, 2000).
9. S. Perlmutter, "Physics Today" 56/4, p. 53 (2003).
10. M. Rees, "Just Six Numbers" (Basic Books: New York, 2000).
11. J. A. Gonzalo "Inflationary Cosmology Revisited", Ibid., p. 41.
12. Letter of Einstein to Lemaitre in John Farrel, "The Day without Yesterday" (Thunder's Mouth Press: New York, 2005) p. 143.

Chapter 19

Concluding Remarks

The Universe has been called "mysterious" (Jeans),[1] "accidental" (Davies),[2] "extravagant" (Kishner),[3] but, perhaps with as much or more reason it can be called "intelligible". Today, astrophysical cosmology at least in part is at least in part capable of putting the universe in numbers.

Let us conclude with a long quotation[4] of George Lemaitre (1894–1966), the Belgian priest who was the father of the Big Bang idea:

"Tous deux — le savant croyant et le non-croyant s'efforcent à déchiffre le palimpseste multiplement imbriqué de la nature où les traces des diverses étapes de la longue évolution du monde se son recouvertes et confondues. Le croyant a peut-être l'avantage de savoir que l'énigme a une solution, que l'écriture sous-jacente est en fin de compte l'œuvre d'un être intelligent, donc le problème posé par la nature a été posé pour être résolu et que sa difficulté est sans doute proporortionnée à la capacité présente ou a venir de l'humanité. Cela non lui donnera peut être pas de nouvelles ressources dans son investigation, mais cela contribuera à l'entretenir dans le sain optimisme sans lequel un effort soutenu ne peut se maintenir long tems ".

Or in my English, with some Spanish flavor:

"Both — the believing and the non- believing scientist — will make efforts to decipher the multiply compound palimpsest of nature, where the traces of the various stages of the long evolution of the world are confused and covered up. The believer may have the advantage of knowing that the enigma has a solution, that the underlying written text is after all, the work of an intelligent being, given that the problem proposed by nature has been proposed to be solved, and that its difficulty is no doubt proportioned to the present or future capacity of men. This will not give him, perhaps, any more resources in his research, but it will contribute to maintain him in that healthy optimism without which no sustained effort lasts for very long".

Bibliography

1. Sir James Jeans, "The Mysterious Universe" (AMS Press, 1933).
2. P. C. W Davies "The Accidental Universe" (Cambridge university Press: Cambridge, 1982).
3. Robert P. Kirshner, "The Extravagant Universe" (Princeton University Press: Princeton, NJ, 2002).
4. O. Godart-M. Heller, "Les relations entre la science et la foi chez Georges Lemaître", in "Pontificia Academiae Scientiarum, Commentarii", Vol. III, n. 21, p. 11. Quoted in the Address of His Holiness John Paul II, the 10th of November 1979 to the Plenary Session of the Pontificia Academiae Scientiarum in commemoration of A. Einstein (Pontificia Academia Scientiarum Scripta Varia: Citta del Vaticano, 1986).

Glossary

Antimatter: Matter consisting of antiprotons, antineutrons and positrons, i.e. of antiparticles which can annihilate with the corresponding normal matter particles, producing pure radiation (photons).

Baryogenesis: Process by which a definite population of baryons (protons, neutrons) becomes well defined in the early universe. This requires that the cosmic matter density becomes equal or less than nuclear density, which occurs at a cosmic temperature $T_b \approx 3.88 \times 10^{12}$ K. It must take place at $t \sim 10^{-5}$ sec or at $t \sim 1$ sec depending on the cosmic equation of state valid at $t(T_b)$.

Baryons: Protons and/or neutrons which make up overwhelmingly ordinary matter since the masses of electrons ($m_e \ll m_b$) and neutrinos ($m_v \ll m_e$) are much smaller than the *masses* of baryons ($m_b \sim 1836\, m_e$).

Big Bang: term coined by Fred Holey in 1950. Holey used it with reference to models of the universe, such as Lemaìtre's and Gamow's that evolved from a super-dense state or singularity at a given instant.

Big Bang nucleosynthesis: Primordial process by which cosmic protons and neutrons fuse together to form ^{4}He nuclei and traces of other light nuclei. This occurs at a temperature $T_{ns} \sim 4.60 \times 10^8$ K. It must take place at $t \sim 10^3$ sec or depending on the equation of state valid at $t(T_{ns})$.

Blackbody radiation: A blackbody is an object which is in thermal equilibrium at a given temperature and emits radiation according to Planck's formula. The intensity of the emitted radiation depends smoothly on frequency at any given fixed temperature and presents a characteristic maximum. (See Chapter 2).

Black hole: A physical system so massive and compact that in it the strong gravitational field prevents even light from escaping. The primordial cosmos could not have been a black hole because in that case the Big Bang could not have taken place.

Blue shift: If a star or galaxy moves towards us the radiation emitted observed from the Earth appears shifted towards shorter wavelength (towards the bine) (See Doppler shift).

Cepheid variable: A type of variable star whose brightness was seen to vary over time. Henrietta Swann Leavitt discovered in 1912 that an approximately linear relationship existed between the period of a Cepheid's pulsation and its luminosity. The longer the period of its pulsation, the greater was the star's luminosity. Once the distance to nearby Cepheids was determined, they became useful measuring sticks for establishing the distances to other galaxies where they were discovered. In 1925 Edwin Hubble discovered a Cepheid in M31, the Andrea Nebula, which enabled him to estimate that it was eight hundred thousand light-years from Earth. It convinced astronomers that Andromeda was not pare of the Milky Way, but a galaxy in its own right.

Chaotic inflationary universe theory: A version of the inflationary universe (Linde, 1983) in which the energy density of the hypothetic inflationary field has a unique minimum at the centre.

Closed universe: A homogeneous and isotropic universe in which the gravity associated to the matter density is strong enough to eventually reverse an initial process of expansion.

COBE: Cosmic Background Explorer, an orbiting satellite with three scientific instruments on board.

Cosmic microwave background radiation: See blackbody radiation. The low-level background noise of radio waves detectable from every direction in space. Its associated equivalent temperature measures about three degrees Kelvin above absolute zero. It was predicted by Gamow, Alpher, and Herman in 1948 as a consequence of the initial super-hot stage of the universe. As the universe expanded, the radiation has cooled over the billions of years to the attenuated microwaves now detectable. Initially discovered accidentally by Penzias and Wilson in 1965.

Cosmic distance ladder: A sequence of calibrated distances to remote galaxies which allows astronomers to extend the range of distance measurements to greater and greater distances.

Cosmic accelerated expansion: Recent observations of magnitude (related to distance) vs. redshift (relate to recession speed) from far away galaxies indicate that these distant galaxies are receding from us (or we are receding from them) relatively faster than expected from the observed motion of nearby galaxies. If we take the local space-time origin at our galaxy now this can be viewed as cosmic *accelerated* expansion, which would require a cosmological constant (see below). But if we tale the local space-time origin behind the surface of "last scattering", at 13.7×10^9 light years from us, the same observations are viewed as cosmic *decelerated* expansion, requiring no cosmological constant (See Chapter 5).

Cosmic parameter (Ω): Dimensionless parameter giving the ratio between actual mass density (ρ_m) and critical mass density (ρ_c), corresponding to cosmic matter moving barely at escape velocity. $\Omega(t) - \rho_m(t)/\rho_c(t)$ is time dependent, and, at present, using local data $\Omega(t_0) = \rho_m(t_0)/\rho_c(t_0) \approx 0.04$, very small in comparison with $\Omega(t_{af}) = 0.99$, corresponding to the time of atom formation ($t_{af} \approx 300,000$ yrs).

Cosmic time parameter (Ht): Cosmic parameter (dimensionless) giving the product of the Hubble's parameter $H = \dot{R}/R$ and the cosmic time at any moment in the cosmic expansion. For an open universe Ht is less than one and more than two thirds (2/3). This parameter is time dependent, because H(t) is time dependent and, of course, t is also time dependent. At present, using local data $H \cdot t \approx 0.9$, substantially larger than $H_{af}t_{af} \approx 0.667 \approx 2/3$ corresponding to the time of atom formation ($t_{af} \approx 300,000$ yrs).

Cosmological constant: A term in Einstein's cosmological equations of general relativity which results in repulsive interaction with opposite sign to the gravitational attraction. Inflationary theories interpret this term as a measure of the energy density of the vacuum. Without the term, Einstein realized his spherical model of the universe would collapse in on itself. Although he introduced the term on the left side of his field equations, essentially to keep the balloon inflated, or "prop up" the geometry of space-time, Lemaitre later reinterpreted Λ (the cosmological constant) as a measure of negative pressure, which he incorporated on the right side of the equations dealing with stress-energy and its effect on the curvature of space-time.

Density perturbations: Fluctuations of the density of matter or radiation in the early universe which can be later amplified by gravity resulting in proto galaxies at carly cosmic times.

DIRBE: Diffuse Infrared Background Experiment. One of the three COBE instruments, cooled at 1.5 K by liquid helium and designed to measure any cosmic primordial infrared radiation as well as any other infrared radiation from stars and dust in the Milky Way and other galaxies. It works at 10 different wavelengths from 1.2 to 240 micrometers.

DMR: Differential Microwave Radiometer. One of the three COBE instruments designed to look for small anisotropies in the cosmic microwave background radiation; it works at 3 wavelengths, 3.3, 5.7 and 9.6 mm. The 3.3 and 5.7 mm receivers operate at 140 K, and the 9.6 mm works at room temperature.

Doppler shift: Shift in the receiver frequency (and wavelength) of wave (sound or electromagnetic radiation) due to the relative motion of source and observant depending on the approach or recession, for light, i.e. electromagnetic radiation, results in a blueshift or a redshift, respectively. Christian Doppler first proposed this effect in terms of pitch for sound waves in 1841, but modern astronomers apply it to electromagnetic waves, corresponding to an observed change in color. Stars or galaxies with spectra shifted to the blue, or shorter, wavelength end of the spectrum are considered approaching the point of view of the Earth. Stars and nebulae with spectra shifted to the red, or longer, wavelengths are considered moving away.

Dust solution: A solution to Einstein's field equations that Lemaître developed in 1932. Instead of a uniform sphere of fluid, whose pressure would go to infinity and thus establish a singularity at the Schwarzschild limit, Lemaître suggested using a sphere of dust, of zero pressure, and was able to show that in fact the singularity at Schwarzschild's limit was only apparent; one could theoretically model a sphere whose radius collapsed to a singularity at radius $r = 0$.

Electromagnetic interactions: Interactions due to electric charges. Static charges give rise to electric fields, while accelerated/decelerated charges produce electromagnetic waves, including radio, microwave, infrared, visible, ultraviolet light and X-rays, as well as y-rays.

Expansion of the universe: The expansion of cosmic space suggested by Einstein's field equations of general relativity. Alexander Friedmann showed in 1922 that such an expansion was a natural consequence of the field equations, but Einstein disagreed and essentially ignored Friedmann (who died in 1925) until Lemaitre independently redeveloped his model more rigorously.

Electroweak interactions: Unified description of weak interactions (responsible for nuclear beta decay) and electromagnetic interactions, due to Glashow, Weinberg and Aldous Salam (1967–1970).

Flat universe: A homogeneous, isotropic universe just at the borderline between spatially *closed* (with a halt in the expansion followed by contraction) and spatially *open* (expanding for ever with non zero acceleration). The geometry for a flat universe is precisely Euclidean. The general relativity cosmological equations are consistent with a flat universe only if the total mass of the universe is infinity, which brings forth the spectre of Olbers gravitational paradox.

Flatness problem: A problem, real or rhetorical, of traditional Big Bang theory pointed out by proponents of cosmic inflation which requires that $k = 0$ ($\Omega = 1$) always, regardless of observational evidence to the contrary in our local neighbourhood ($R \sim R_0$, $t \sim t_0$).

Gauge theories: Theories which allow transformations in the dynamical equations leaving invariant the scalar and vector potentials describing physical interactions. The theoretical equations describing electromagnetic interactions (QED) are an example of gauge theory. The theoretical equations describing the strong nuclear forces, i.e. quantum chromodynamics (QCD) equations, are an example of gauge or Yang-Mills theory resulting in the so called asymptotic freedom for the way quarks are bound within the nucleons.

General theory of relativity: Einstein's theory of universal gravitation, which describes gravity in geometric terms, as a property of space-time. The geometry of space, or the curvature of space-time, is determined by the presence of mass and energy. Among its first known consequences, the theory predicted that light would bend in the presence of massive objects and that gravity would also slow clocks in the field of massive objects. As a geometric theory, general relativity allowed Einstein to model the entire universe as the totality of all existing matter. This opened up an entire new field of cosmology as an exact science in the twentieth century.

Gluons: Particles which play a role in the strong interactions analogous to that of photons in electromagnetic interactions.

Grand unified theories: A speculative class of theories of particle interactions which attempts a unified description of the electromagnetic, weak and strong interactions, leaving aside the gravitational interaction.

Gravitation: The mutual attraction between any two massive particles. In a cosmic scale gravity appears as strong because it has an infinite range, like the electromagnetic interaction, and unlike the weak and strong nuclear forces, but is always attractive. Only in inflationary theory a cosmic false vacuum is postulated that results in a repulsive force resembling a negative gravitation.

Homogeneity: An attribute assumed for early cosmological models, the assumption that matter is distributed evenly throughout the volume of the universe.

Horizon problem: A problem, real or rhetorical, of traditional Big Bang cosmology pointed out by proponents of cosmic inflation. If at very early times after the Big Bang the universe was homogeneous and isotropic to start with, the horizon problem disappears altogether.

Hubble parameter: The ratio of recession velocity of galaxies (R) to distance (R), improperly called Hubble's constant, because for nearby galaxies $H_0 \approx \dot{R}_0/R_0$ is approximately constant, but, in principle, it is time dependent, being $H(t) \approx \dot{R}(t)/R(t) >> H(t_0)$ for early times. The presently accepted value of the Hubble parameter is $H_0 \sim 67$ (km/s)/Mpc, being 1 Mpc (Megaparsec) = 3.26×10^6 lightyears.

Hubble time: Estimate for the age of the universe based on the reciprocal of Hubble's constant. At the time Lemaître originally proposed his primeval atom theory, the Hubble time was only estimated to be only about two billion years. Later revisions of Hubble's distance estimates increased the value upward to four billion years in 1948 and upward to ten billion by the time Allan Sandage took over Hubble's position at Mount Palomar.

Inflationary cosmology: The cosmological theory which assumes a very early ($\sim 10^{-35}$ sec after the Big Bang) exponential expansion *at constant density* during which the total mass of the universe increases by many orders of magnitude. In the inflationary process the new matter comes out of the vacuum itself. The process is somewhat analogous to the continuous creation of matter out of nothing in the steady-state-cosmology of Bondi, Gold and Hoyle, which was popular in the 50's and early 60's, before the cosmic background microwave radiation was discovered by Penzias and Wilson in 1965.

Ionised atoms: Atoms under physical conditions (f.i. high radiation pressure) such that one or more electrons have been taken apart. At times below 300.000 years and temperatures higher than 3880 K the universe consisted in a *plasma* of ionised atoms, mainly hydrogen (76%) and a substantial amount of ^{4}He (23%). Later, when the temperature was cooling progressively, the universe begun to consist in a gas of neutral atoms. Substantially later, the gravitational interaction began to pull atoms together to form protostars within protogalaxies. Finally, the universe evolved towards its present, slowly changing, overall physical conditions.

Isotropy: Attribute assumed Lemaître, and other cosmologists – that the universe appears the same in all directions. To a very high degree, measurements of the cosmic background radiation bear this out.

Leptons: A class of non-strongly interacting particles which includes the electron, the muon, the tau particle, and their associated neutrinos (See Chapter 3).

Local group: A group formed by about twenty galaxies surrounding our Via Lactea (Milky Way).

Local supercluster: An assembly of about 100 clusters of galaxies including our local group.

Magnetic monopole: An isolated pole (north or south) of a hypotetic magnet. An infinitely long magnet in which the two poles are so far apart than they do not affect each other would act as a pair of monopoles. Dirac predicted the existence of particles with properties of magnetic monopoles, and so did some grand unified theories. But they have never been observed.

Magnetic monopole problem: A problem, real or rhetorical, of traditional Big Bang cosmology originally pointed out by the proponents of cosmic inflation. Since the observation of real magnetic monopoles has not been confirmed, this problem appears to have faded away.

Meson: A strongly interacting particle consistent in a quark and an antiquark.

Microwave: An electromagnetic wave with wavelength between one millimetre and about 30 centimetres.

Neutrino: An elementary particle of very small rest mass which is electrically neutral and very weakly interacting with other material particles. It was predicted in 1931 by Pauli and detected experimentally in 1956 by Cowan and Raines.

Observed cosmic mass: Einstein's cosmic general relativistic equations are consistent with a very large but finite mass for the entire universe. Present cosmic dynamics allows one to estimate the total cosmic mass as $M_u \sim 4.5 \times 10^{54}$ g. Note that this number if of the order of $N_G N_s M_{sun}$, where $N_G \sim 10^{11}$ (galaxies) times $N_s \sim 10^{11}$ (stars) times $M_{sun} \sim 2 \times 10^{33}$ g, the mass of a typical star (the Sun).

Open universe: A homogeneous isotropic universe in which gravity is not strong enough to stop the expansion. So an open universe goes on expanding for ever. Local cosmic dynamics data support an open universe with a cosmic space curvature $k < 0$.

Phase transition: A sudden change in the properties of a material system produced by varying temperature, pressure or other physical parameter. Inflationary theory postulates a cosmic phase transition at about 10^{-37} seconds after the Big Bang.

Photon: A quantum (discrete) minimum of electromagnetic energy consisting essentially in a localized particle of light añer the quantum theory of radiation put forward by Max Planck in 1900.

Planck's time: A "natural" unit of time proposed by Max Plank in terms of universal constants: $t_{pl} \approx (\hbar G/c^5)^{1/2} \approx 5.4 \times 10^{-44}$ sec.

Primeval atom (*l'hypothese de l'atom primitif*): The name Lemaître gave to his theory of the universe initiating from a hyper-dense sphere of cold matter, disintegrating via radioactivity into an expanding universe. His theory was also more accurately described as "the cosmic egg."

Quark: Elementary constituents of protons, neutrons and other strongly interacting particles (baryons and mesons).

Red giant: A phase in the life cycle of a typical star (not very heavy, like our Sun) in which the size of the stars increases tremendously and then blows up into space, leaving as a remnant a white dwarf.

Redshift: Shift towards longer wavelength (smaller frequency) in the light emitted by stars or galaxies moving away from us (See Doppler Shift).

Relativity: The special theory of relativity, proposed by Albert Einstein in 1905, assumes that all light propagates at a constant (invariant) speed c in free space, regardless of the relative motion of source and observant. The General Theory of Relativity, produced by Einstein in 1915, constitutes a general theory of gravity consistent with special relativity, and therefore, with the invariance of c. It predicts that light rays are deflected in their path when coming close to large masses. The observation of this effect taking advantage of a Solar eclipse in an expedition leaded by Eddington in 1919, did make Einstein instantly famous all over the world, and confirmed the theory of relativity.

Renormalization: A theory such as quantum electrodynamics (QED) which in first approximation gives answers in agreement with experiment, in a second approximation, however, may give divergent answers (infinity) which are unrealistic. Renormalization is a mathematical transformation on the theoretical equations to avoid the unwanted infinities.

Schwarzchild solution: The first complete solution to Einstein's field equations of general relativity, proposed by German astronomer Karl Schwarzschild shortly before he died in 1916. Schwarzschild assumed that for a uniform, fluid sphere, a singularity would occur as the spheres radius shrank to $r = 2GM/c^2$, where G is the gravitational constant, c is the speed of light, and M is the mass of the sphere.

Singularity: When the standard Big Bang theory is extrapolated back to time zero, a number of cosmic physical parameters become infinity, f.i. the density, the pressure, the temperature. It is physically meaningful to tray to investigate what happens as t → 0 and R → 0, but there is always a limit, imposed by Heisenberg uncertainty principle which sets a minimum conceivable cosmic time ($t_H \approx \hbar/M_u c^2 \approx 2.8 \times 10^{-103}$ sec) beyond which it is pointless to speculate within the realm of physical theory.

Special theory of relativity: Einstein's theory combines two principles - that the speed of light is constant, irrespective of the velocity of its source, and that the laws of physics hold universally for all inertial frames of reference, meaning coordinate systems at rest or moving with a constant velocity. Consequence of the theory are (1) that observers in different reference frames moving relative to one another will record different results for the time of events in a third frame; and (2) that energy is equal to mass multiplied by the speed of light squared.

Standard model of particle physics: A general theory of particle interactions which describes successfully the electromagnetic, the weak and (to a large extent) the strong interactions. This theory was developed in the early 1970's and leaves out the gravitational interaction.

Steadily driven expansion (SDE): An alternative to inflationary theory based upon the traditional Big Bang theory in which the large increase in cosmic radius from very early cosmic times ($t_H \approx 2.8 \times 10^{-103}$ sec) to times beyond Planck's time ($t_{pl} \approx 5.4 \times 10^{-44}$ sec) takes place steadily instead of instantly.

Steady state theory: A cosmological theory postulated by Fred Hoyle, Thomas Gold, and Herman Bondi in Britain in 1948, reasserting that the universe is in essence without change. Hoyle, Bondi, and Gold argued against a temporal, Big Bang origin, suggesting that while the universe was expanding, this occurred in a natural state similar to that suggested by de Sitter's universe, but with matter being constantly created in the form of hydrogen atoms in empty space to sustain the existence of the cosmos in perpetua. In the early 1960s the discovery of quasars and other developments in astronomy strengthened the contention that the universe in the distant past was far different than at present. The discovery of the cosmic microwave background radiation all but discredited the theory.

Strong interactions: The interactions which bind together quarks to form protons, neutrons, etc.

Thermal equilibrium: Physical conditions which imply that photons (radiation) have a characteristic spectral distribution, the so called blackbody spectral distribution. COBE confirmed that cosmic thermal equilibrium prevailed at the time of atom formation (T ~ 3880 K), and presumably much earlier, f.i. at nucleosynthesis, baryon formation, Plank epoch, and even before.

Type Ia supernovae: Violent stellar explosions, signaling the utter demise of a star, which take place when a white dwarf bleeds enough mass from a companion star to pass a certain mass limit, approximately 1.4 times the mass of the sun (also called Chandrasekhar's Limit after the Indian-American astrophysicist Subrahmanyan Chandrasekhar), at which point it explodes. The luminance of such supernovae is so great that they can be brighter than the entire galaxies in which they are photographed. The consistent level of luminosity of Type Ia supernovae have made them an even more reliable distance indicator than Cepheid for determining extragalactic distances.

Type II supernovae: Explosions of stars of masses at least eight times that of the sun. Having reached a state where all nuclear fusion has reduced the star's core to iron and heavier elements, the star can no longer sustain fusion to provide the energy necessary to counteract the force of gravitational collapse. The core contracts into a neutron core or black hole, while the outer layers of lighter matter are blown off into space at thousands of kilometers per second.

Uncertainty time: The earliest conceivable cosmic time allowed to be considered by the uncertainty principle ($t_H \approx 2.8 \times 10^{-103}$ sec).

Weak interactions: Short range nuclear interactions responsible for beta decay. Neutrinos are subject only to this type of interactions.

White dwarf: Final phase in the Ufe of a typical star after it went through the phase of red giant.

Ylem: The primordial cosmic material after Gamov and collaborators. When the cosmic density was incredibly high, matter density, which then could have been present in an amount comparable to that of radiation or not, was so incredibly high that far exceeded that in ordinary baryons or electrons.

Index